OBSTACLE AVOIDANCE IN MULTI-ROBOT SYSTEMS

Experiments in Parallel Genetic Algorithms

WORLD SCIENTIFIC SERIES IN ROBOTICS AND INTELLIGENT SYSTEMS

Vol. 1: Genetic Algorithms and Robotics — A Heuristic Strategy for Optimization (*Y Davidor*)

Vol. 2: Parallel Computation Systems for Robotics: Algorithms and Architectures (*Eds. A Bejczy and A Fijany*)

Vol. 3: Intelligent Robotic Planning Systems (*P C-Y Sheu and Q Xue*)

Vol. 4: Computer Vision, Models and Inspection (*A D Marshall and R R Martin*)

Vol. 5: Advanced Tactile Sensing for Robotics (*Ed. H R Nicholls*)

Vol. 6: Intelligent Control: Aspects of Fuzzy Logic and Neural Nets (*C J Harris, C G Moore and M Brown*)

Vol. 7: Visual Servoing: Real-Time Control of Robot Manipulators Based on Visual Sensory Feedback (*Ed. K Hashimoto*)

Vol. 8: Modelling and Simulation of Robot Manipulators: A Parallel Processing Approach (*A Y Zomaya*)

Vol. 9: Advanced Guided Vehicles — Aspects of the Oxford AGV Project (*Eds. S Cameron and P Probert*)

Vol. 10: Cellular Robotics and Micro Robotic Systems (*T Fukuda and T Ueyama*)

Vol. 11: Recent Trends in Mobile Robots (*Ed. Y F Zheng*)

Vol. 12: Intelligent Assembly Systems (*Eds. M Lee and J J Rowland*)

Vol. 14: Intelligent Supervisory Control: A Qualitative Bond Graph Reasoning Approach (*H Wang and D A Linkens*)

Vol. 15: Neural Adaptive Control Technology (*Eds. R Zbikowski and K J Hunt*)

Vol. 17: Applications of Neural Adaptive Control Technology (*Eds. J Kalkkuhl, K J Hunt, R Zbikowski and A Dzielinski*)

Forthcoming:

Vol. 13: Sensor Modelling, Design and Data Processing in Confined Environments (*M D Adams*)

Vol. 16: Advances in Robotics & Automation for Hazardous Environment (*Eds. P Lever and F Y Wang*)

Vol. 18: Soft Computing in Systems and Control Technology (*Ed. S Tzafestas*)

Vol. 19: Adaptive Neural Control of Robotic Manipulators (*S S Ge, T H Lee & C J Harris*)

Series in Robotics and Intelligent Systems – Vol. 20

OBSTACLE AVOIDANCE IN MULTI-ROBOT SYSTEMS

Experiments in Parallel Genetic Algorithms

MARK A C GILL & ALBERT Y ZOMAYA
University of Western Australia

World Scientific
Singapore • New Jersey • London • Hong Kong

Published by

World Scientific Publishing Co. Pte. Ltd.

P O Box 128, Farrer Road, Singapore 912805

USA office: Suite 1B, 1060 Main Street, River Edge, NJ 07661

UK office: 57 Shelton Street, Covent Garden, London WC2H 9HE

Library of Congress Cataloging-in-Publication Data

Gill, Mark. A. C.

Obstacle avoidance in multi-robot systems : experiments in parallel genetic algorithms / Mark A. C. Gill, Albert Y. Zomaya.

p. cm. -- (World Scientific series in robotics and intelligent systems ; vol. 20)

Includes bibliographical references and index.

ISBN 9810234236

1. Robots -- Control systems. 2. Genetic algorithms. 3. Parallel processing (Electronic computers) 4. Intelligent control systems. I. Zomaya, Albert Y. II. Title. III. Series.

TJ211.35.G55 1998

629.8'925275 -- dc21 97-52397

CIP

British Library Cataloguing-in-Publication Data

A catalogue record for this book is available from the British Library.

This book is printed on acid-free paper.

Printed in Singapore by Uto-Print

PREFACE

Future robots are expected to achieve higher degrees of autonomy, which requires massive computing power. Moreover, a robot's computer system should be scalable in that it can be expanded to future demands without sacrificing performance. One attractive solution to this problem is to use parallel computing. Nowadays high performance computer system can be built by using inexpensive commodity components. These systems can be employed to compute a wide range of robotic algorithms *in situ*.

A parallel computer is one that consists of a collection of processing units that cooperate to solve a given problem by working simultaneously on different parts of that problem. In principle, there is no limit to the number of actions that can be executed in parallel, thus, offering an arbitrary degree of improvement in computing speed, especially, when compared to traditional uniprocessors. This book examines the use of parallel computing techniques combined with genetic algorithms to propose effective solutions to the path planning problem for robot manipulators.

Robots belong to a class of real-time systems in which severe time constraints have to be met. Ideally, a computer system attached to a robot should provide sufficient computing power to enable the implementation of a wide range of complex algorithms that are normally needed for the proper functioning of the whole system (robot, computer, sensors, etc.). Robotic algorithms cover a wide area ranging from mechanical manipulation to computer vision and artificial intelligence, however, in this book we will restrict ourselves to investigating path planning problems which have theoretical importance as well as practical implications. In addition, the efficient solution of the path planning problem should influence the development of algorithms in other areas of robotics such as control and vision.

The path planning problem has been addressed by many researchers, but in this book we investigate the use of genetic algorithms in solving this problem. Genetic algorithms belong to a wide class of stochastic optimization algorithms generically known as Evolutionary Computing, which are based on principles that originated from studies on organic evolution. We decided to base our study on genetic algorithms because among other evolutionary computing paradigms they seem to be the most robust and straightforward to apply.

Normally, optimization techniques tend to be compute-intensive and genetic algorithms are no exception. However, genetic algorithms, as it is the case with many biologically-inspired computational paradigms (e.g. neural networks, cellular automata), are inherently parallel. This makes them suitable candidates for parallel computing implementations. By combining the robustness of genetic algorithms with the high performance delivered by parallel computers, we can develop powerful optimization algorithms that can solve formidable problems at a very fast rate. This is the approach upon which the ideas for tackling the path planning problem as presented in this book were built. There are other techniques that were proposed in the literature over the last few years to solve the path planning problem, nevertheless, we hope that this book will serve a purpose.

The development of parallel algorithms can be realized through several levels. One level, which we adopted in this work, is to focus on the details of implementation on a particular machine or architecture. This approach allows for a better understanding of the computational and communication complexities of a given implementation. Parallel computers are driven by technology. Everyday new devices are introduced adding to the existing capabilities. The key issue is to learn about parallel computers in the context of today's technology. It is important to look at both cost and performance when evaluating an architecture or an implementation and not only performance.

The material in the book is interdisciplinary in nature. It combines topics from certain aspects of robotics, genetic algorithms, and parallel processing. The book can be used by senior undergraduates, graduates, and researchers in Electrical and Computer Engineering, and Computer Science.

Acknowledgements

Many people have contributed in various ways to the compilation of this book. We would like to express our thanks and deepest appreciation for the members of the Parallel Computing Research Lab at the Electrical and Electronic Engineering Department, The University of Western Australia.

During this work, we have had the benefit of direct and indirect comments and suggestions from many talented people. Some of them, as often happens, may not even be aware of the extent of influence they had on us, and on the work. We would like to thank Professor Stephan Olariu (Old Dominion University), and Professor Ivan Stojmenovic (Ottawa University), and Professor Mounir Hamdi (University of Science and Technology, Hong Kong) for doing an excellent job at reviewing the early drafts of the manuscript and for their valuable comments that helped in improving the quality of the final product.

We also would like to extend our thanks to Professor Chris Harris (Southampton University) for his support of this book. Finally, many thanks go to our families for their help, support, and inspiration.

Mark A. C. Gill
Albert Y. Zomaya

Perth, Western Australia
November, 1997

ABBREVIATIONS

Symbol	Definition
CPU	Central Processing Unit
CSG	Constructive Solid Geometry
DH	Denavit-Hartenberg
DOF	Degrees of Freedom
GA	Genetic Algorithm
IC	Integrated Circuit
LM-SIMD	Local Memory, Single Instruction stream, Multiple Data stream
LSI	Large Scale Integration
MFLOPS	Million of FLOating Point operations per second
MIMD	Multiple Instruction stream, Multiple Data stream
MIPS	Million of Instructions per Second
MISD	Multiple Instruction stream, Single Data stream
MSI	Medium Scale Integration
PGA	Parallel Genetic Algorithm
PVM	Parallel Virtual Machine
SIMD	Single Instruction stream, Multiple Data stream
SISD	Single Instruction stream, Single Data stream
SM-SIMD	Shared Memory, Single Instruction stream, Multiple Data stream
SPMD	Single Program, Multiple Data
SSI	Small Scale Integration
TRAM	TRAnsputer Module
VLSI	Very Large Scale Integration

CONTENTS

CHAPTER 1

OVERVIEW

1.1 Introduction

Multiple robotic systems are becoming more common with more than one robot working in the same environment, such as today's automated factories, where there is an ever increasing demand for flexible automation. Collisions occurring between the robots and collisions with objects in the environment are possible. There currently exists a lot of research into improving robots to enable them to do more complex tasks with less human intervention. Part of this research is centered around path planning. Path planning is devising collision free trajectories for each robot manipulator to enable them to move about appropriately and to perform their required tasks. A collision can be expensive in terms of both the cost of repair and the time taken to perform the repair.

Path planning is generally performed off-line. That is, the robot or robots are shut down and a new trajectory or trajectories are created and tested before being used. These manipulators operate in highly structured and controlled environments so all expected obstacles are known in advance. It also allows the optimal trajectory (in terms of minimum time or energy) to be found. The robot itself does not participate in the planning process - it only follows the new trajectory when switched back on-line.

1.2 Robotics

A robot is a *reprogrammable multifunctional manipulator designed to move materials, parts, tools, or specialized devices through variable programmed motions for the performance of a variety of tasks* (Rehg 1985). Or, another definition is that a robot is *a software-controllable man-made device that uses sensors to guide itself and/or its end-effector through deterministic motions in order to manipulate physical objects* (Shilling 1990). The robot is designed to be versatile and to function automatically without human intervention while performing a set task. When the task is to be changed the robot can be reprogrammed to perform a new task without expensive retooling.

Robots were initially developed over twenty years ago and evolved out of the teleoperator and numerically controlled machines. The teleoperator is a mechanical arm remotely controlled by someone at a distance. The numerically controlled machine is a milling machine which receives its commands via a tape reader. Changing its operation was merely a matter of reprogramming it with new tape. Initially robots were used in uniform pick-and-place operations. Gradually, as the technology developed, the tasks have become more complex and less constrained. The development of robot manipulators (i.e. the mechanical arm, which is the most common form of the robot) is tied to the development of digital computers, since they are used to control manipulators.

Robots are being used more widely and their number is constantly increasing as they become more economically viable, more flexible and advances in computer technology make them more intelligent. The use of robots in industry has several advantages: (1) Increased productivity due to their speed and ability to work around the clock and therefore better utilization of capital equipment; (2) Improved product quality by repeating the same task consistently; (3) Flexibility with the ability to be reprogrammed to perform a new task; and (4) Precision by being able to move accurately (Kafrissen and Stephans 1984).

The robot manipulator can be classified in various ways (Rehg 1985; Kafrissen and Stephans, 1984; Schilling 1990; Critchlow 1985; Klafter *et al* 1989):

- Work envelope geometry or mechanical configuration - The basic shape of the workspace in which the manipulator operates. The geometrical configuration of the manipulator will determine the shape of the workspace. The base of the manipulator may be mobile or fixed. Most manipulators fall into the following categories: (1) Cartesian or rectangular manipulators; (2) Cylindrical manipulators; (3) Spherical or polar manipulators; (4) Revolute or articulated manipulators (either horizontal or vertical); and (5) Snakelike or tensor-arm manipulators (Shahinpoor 1987; Yoshikawa 1990).

- Drive System - The mechanism used to power the joints of the manipulator, such as: (1) Hydraulic; (2) Pneumatic; (3) Electric; or (4) A combination of the first three.

- Motion control - The method that the robot controller uses to guide the end-effector (robot hand or tool) of the manipulator through space. The most common techniques are: (1) Stop-to-stop; (2) Point-to-point motion; (3) Continuous path, and (4) Controlled path.

- Capability level or intelligence - The relative intelligence of the robot determines the complexity of the task it is able to perform and under what conditions it is able to perform the operation, ranging from low and medium through to high. The capability levels are: (1) Sequence-controlled machines; (2) Playback machines; (3) Computed-trajectory machines; (4) Controlled-path robots; (5) Adaptive robots; and (6) Intelligent robots.

- Applications - The type of work that the robot has been designed to do. Some of the main broad categories are: (1) Automated or flexible manufacturing; (2) Remote exploration; (3) Prosthetic, exoskeletons and biomedical; (4) Hazardous material handling; (5) Locomotive mechanisms; and (6) Service (Shahinpoor 1987; Klafter *et al* 1989).

The above categories are fairly broad and robots can be further characterized by more specific parameters, such as: (1) Number of axes; (2) Load carrying capacity; (3) Maximum speed or cycle time; (4) Reach and stroke; (5) Tool orientation; (6) Repeatability; (7) Precision and accuracy; and (8) Operating environment.

A robot system generally consist of three components: (1) The motion system, which is the physical structure; (2) The recognition system, which uses various sensors to gather information about its environment (external sensors, such as tactile, vision, proximity and range sensors) and its configuration (internal sensors, such as optical encoders, potentiometers, tachometers, strain gauges, and microswitches); and (3) The control system, which uses the information from the recognition system in conjunction with the motion system to perform a desired task (Yoshikawa 1990). Two types of control can be performed: (1) Position control deals with moving the end-effector along a desired trajectory; and (2) Force control, used to make the end-effector exert a desired force on a particular load.

1.3 Path Planning, Genetic Algorithms and Parallel Processing

Path planning is an important aspect of robotics research. It is part of the motion planning process for manipulators. Without path planners the motions of manipulators would need to be constantly specified exactly by the operator. The main obstacle in path planning is the complexity of the problem. It is NP-complete (Hwang and Ahuja 1992), and therefore to find a solution in reasonable time a heuristic algorithm must be used. The time can be further reduced by taking advantage of parallel processing which can speed up the execution of an algorithm by executing different tasks simultaneously. Parallel processing has recently begun to appear in solving robotics problems and is becoming a viable solution as the cost of hardware is dropping and is becoming more advanced.

The GA (Genetic Algorithm) is an unorthodox technique (in the way it performs its search) which is amenable to parallel processing implementations due to its inherent parallel nature. It is not guaranteed to produce an optimal solution, but can often solve problems rapidly to within a specified accuracy. It is highly suited to solving path planning problems where a reasonable solution is required quickly. Hwang and Ahuja (1992) stated that practical algorithms to solve the general motion planning problem have not been developed.

1.4 Motivation for this Work

Robot manipulators have several severe limitations. They usually do not participate in the path planning process and, therefore, the path planning must be done for them off-line. They have a limited ability (if any) to learn from past experiences and to respond intelligently to unexpected changes in their environment. Robots can presently be fitted with sensors that allow them to perceive some changes in their environment, but often cannot modify their motion to react to any unforeseen changes while still carrying out their task. Some incorporate only a very basic ability (such as stopping if they detect that they have hit something - this feature is an essential safety precaution in some places where robot manipulators work in the same environment as people). Brady (1989) identified motion planning as one of the problem areas in robot development.

The motion planning problem can be decomposed into several specific areas, including problems with the sensors used (this is out of the scope of this work) and computer planning and control (which have problems with the speed at which data is processed and the type of algorithms used). Motion planning does not just involve creating collision free paths, but also translating these paths into a suitable trajectory for the robot manipulator, and then generating the control signals for the manipulator.

One of the goals of motion planning is to develop robotic systems that overcome the current poor ability of a robot to interact intelligently with an unstructured environment. Part of this involves creating a real-time autonomous motion planner, that is, a robot capable of planning its own course of action on-line without human intervention. The operator should specify the task to be performed rather than how to perform it. Specifying how to perform the task can be very time consuming. The problem is not simple - once a suitable path is found it must be translated into a suitable trajectory for the manipulator. The path is the physical path traced out by the end-effector in Cartesian space, and the trajectory is the set of joint variable motions which enable the end-effector to follow the path.

The complexity of this is dependent on the number of robot manipulator links and their interaction. The problem is further complicated when there is more than one manipulator present with overlapping workspaces (multimovers problem) and each manipulator has its own separate task to perform. If the motion planning is performed on-line, then the robot manipulator must be not only able to sense when it is in danger of colliding, but be able to react appropriately to this situation while carrying out its task. A problem with performing on-line planning is that the trajectory that the manipulator follows may not be optimal - it just uses the first suitable one found.

1.5 Contribution of this Work

In this work, the collision detection problem is considered first. Each manipulator and object in the workspace is represented by a mathematical model. The manipulators are treated as a series of connected line segments with an associated thickness, while the objects are treated as either spheres or as a collection of plane segments. The collision detection routine calculates the minimum distances between each manipulator and obstacles (other manipulators and objects) in the workspace. The objects may be moving, stationary or even change shape, and the base of the manipulator can be mobile. If the distance calculated is less than a particular threshold (the size of the link involved) then a collision has occurred.

In the collision avoidance problem, the end-effector of each manipulator is guided iteratively towards the goal and away from obstacles using potential fields or an approximate cell decomposition technique with a greedy depth-first search algorithm. A GA is used to search for suitable joint variables for each manipulator (i.e. generate the trajectory from the path) which ensures that no collision occurs with other obstacles and that the end-effector moves close enough to the desired target during each iteration. The GA has been implemented using a parallel architecture (transputers) to speed up the processing, since the problem is very computationally expensive.

The collision avoidance algorithm is an on-line scheme which uses local knowledge and operates in cartesian space, requiring no mapping into configuration space. The configuration space transformation requires complex calculations and must be continually updated in a time-varying environment. The time taken to perform the configuration space transformation calculations and the requirement to have complete knowledge about the workspace makes it unsuitable for an on-line collision avoidance scheme. The algorithm is also not limited by the number or type of robot manipulators and obstacles. The collision avoidance problem is a more difficult problem to solve than the collision detection problem, and the level of difficulty in both cases depends on the manipulators and the environment.

1.6 Work Organization

Chapter (2) contains a brief introduction to parallel processing and how it has been used in this work. It reviews the basic concepts of parallel processing along with some of the parallel and multiprocessor network classifications. The transputers are introduced along with how they have been employed.

Chapter (3) presents an overview of path planning. The main problems and algorithms are classified and examined. The stages involved in motion planning are covered along with the search techniques used.

Chapter (4) contains an outline of the GA and how it has been implemented on parallel architectures. It also contains a brief summary of other related search methods. The main operators are explained with an example to demonstrate how it works. The implementation on a transputer network is shown.

Chapter (5) illustrates how a GA can solve the inverse kinematics problem. The inverse kinematics problem is described along with the difficulty associated with solving it by inverting the forward dynamics equations.

Chapter (6) presents a collision detection algorithm. Past techniques, both simple and more complex are examined. The manipulators and objects are modelled and the method used to compute distances between the manipulators and obstacles is specified. There are also two examples demonstrating the method.

Chapter (7) demonstrates the collision avoidance algorithm. Initially moving a single point through space is considered. This is then built up upon with a modified GA solution to the inverse kinematics problem coupled with the collision detection routine to form the collision avoidance algorithm. The GA is well suited to parallelization and this is exploited to greatly speed up the collision avoidance algorithm.

Chapter (8) presents several serial examples and a parallel example to demonstrate the algorithm working. In the parallel case a comparison is given to show the speed-up.

Chapter (9) concludes the work. Conclusions are drawn from the results obtained. Limitations and future directions are also discussed. The appendices (1 and 2) provide some supplementary material.

CHAPTER 2

PARALLEL COMPUTING

2.1 Introduction

Over the last few decades computing power has grown enormously. There is a lot of the research around the world devoted to building faster and more powerful computers, and utilising these new machines more effectively. Advances in VLSI technology have produced very fast and powerful processors, but there are finite limits as to how far one can go with this technology. These limits are dependent on the physical properties of the materials used in manufacturing the microchips. One way to overcome this drawback is to identify and exploit inherent parallelism in the task to be performed, in order to implement it on a parallel machine. This enables parts of the task to be performed simultaneously and, therefore, faster.

The earliest electronic computer was developed in 1945 and up until 1954 computers were based on vacuum tubes and relays. From 1955 to 1964 advances in electronics enabled diodes, transistors and magnetic ferrite cores to be used. After the IC was developed, computers built during 1965 to 1974 were constructed using SSI technology and MSI technology ICs. From 1975 up to the present, computers have been manufactured using LSI and VLSI technology. While the level of technology of the hardware has progressively increased, likewise, the sophistication of the software has progressed from data processing to information processing, to knowledge processing and finally through to intelligence processing. The increasing complexity of the software and the constant demand for more power has been instrumental in pushing the hardware development.

There are two limiting factors in the development of faster computers. The first, mentioned earlier, is the physical limits of the material used in the construction of ICs. The speed of light is finite and therefore the communication between various components on the IC is limited. In addition, there is also a limit on the minimum distance between the various components before they begin to interact, affecting their speed and reliability (Akl 1989). The other limiting factor is processor design. Most processors use the von Neumann architecture. This architecture is limited mainly by the CPU-to-memory bottleneck, due to the difference between high CPU speeds and relatively low memory access speeds (Stone 1987).

This chapter will give a brief background of parallel processing and how it is used in this work. Section (2) presents an overview of parallel processing, the ways in which various implementations are classified, network topologies, the main measures of performance, parallel algorithms, and a discussion of real-time systems. In Section (3) the approach adopted in this work is introduced along with the implementation on transputers (Section 4).

2.2 Parallel Processing

Most computers today follow the von Neumann model, as shown in Figure (2.1). The control unit fetches instructions and data from memory, processes the instruction and then returns the result back to memory (Zomaya 1996). There is only one unit of each kind, only one instruction at a time is executed (or one part of an instruction if pipelining is used) on this type of machine and the control is centralised. Various techniques have been employed to speed-up their operation. They are improving the algorithms to better suit a particular machine, improving the hardware (such as cache memory, pipelining and multitasking - essentially still sequential processing), and parallel processing.

Cache memory is high speed memory that holds part of a program or data, near the CPU, that it is expecting to use, to speed-up access times (Baron and Higbie 1992). In pipelining, the CPU does not wait until it has finished executing an instruction before beginning the execution of the next instruction. All instructions go through several stages to be processed and these stages can be overlapped. While an instruction is at one stage, the subsequent instructions are being processed in earlier stages. This allows maximum utilisation of the CPU's instruction processing. Data can also be pipelined - this is typically done on vector processors (Hennessy and Patterson 1994). Multitasking is the ability of a processor to switch between tasks. While one task is idle (e.g. waiting for an event) another can be running (e.g. print spooling). Multitasking can simulate parallel processing by switching rapidly between tasks so they appear to be running simultaneously.

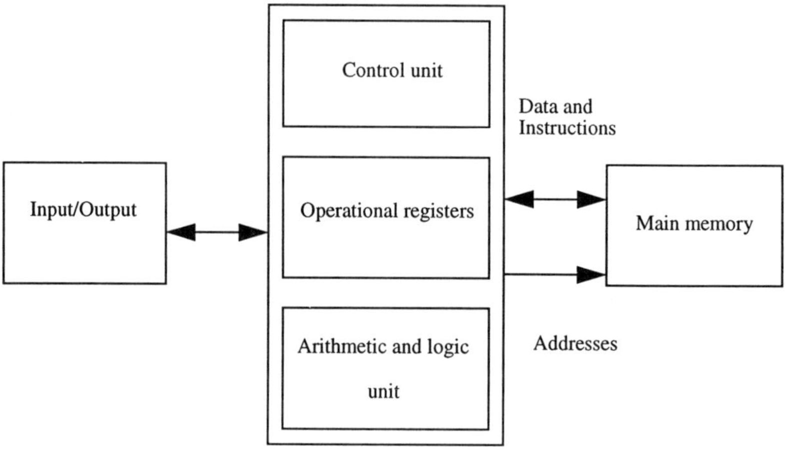

Figure 2.1: The von Neumann model

Parallelism is the process of performing tasks concurrently (Lewis and El-Rewini 1992), i.e. a collection of processing units cooperate together to solve a problem by working simultaneously on different parts of the problem. If several subtasks (processes) can be performed simultaneously then the time taken to complete the overall task can be reduced. The level of reduction depends on how suitable the task is to parallelisation, how well it has been parallelised, the particular architecture used and the number of parallel processing units. A process is a subprogram that executes sequentially but can run in parallel with other processes.

Parallel processing offers several advantages over the traditional sequential processing: (1) It may be more natural to implement a task on a parallel architecture (such as a GA); (2) It may improve the performance of the system (e.g. the speed-up); (3) It will improve the reliability and fault tolerance of the system (if one processor fails then others may be able to take on that one's functions); and (4) Economic factors may suggest that a parallel machine is more cost effective than a high-performance sequential machine (the main criteria used for judging an architecture are its performance, cost, and memory (Stone 1987)). There is an increasing usage of parallel systems and the growth of networks allows a greater use of distributed systems (where the resources are not all centralised in the same location) and cluster computing (where groups of networked computers cooperate together to collectively solve a problem) (Zomaya 1996).

In order to effectively utilise parallelism, a parallel algorithm must be designed to take the best possible advantage of the system on which it is to be implemented. The program may either be written in a parallel language by the programmer, transformed from a sequential program by a special parallel compiler which can identify segments that can run in parallel, or the hardware may utilise any inherent parallelism in the instructions to optimise performance. A parallel algorithm is a solution for a particular problem to be performed on a parallel machine. The way in which a task is implemented depends on: (1) The number of processors in the system; (2) How the processors are connected to each other (to communicate), i.e. the interconnection network; (3) The arrangement of the memory in the system (does each processor have its own memory? Or, do they share memory from a common pool?); and (4) The synchronization between the subtasks (this must be efficiently handled to avoid unnecessary delays and deadlock in the system).

Parallel processing has been used in various applications. Robotics and its related applications are a prime example. Examples include Cela *et al.* (1991), Cheng *et al.* (1992), Lozano-Perez and O'Donnell (1991), Prassler and Milios (1990), Zomaya (1992), and Zomaya (1996). Robotic systems are real-time systems that have hard real-time constraints to be met, they require a very high sampling rate to operate effectively. Refer to Section (2.2.5) for the definition of real-time and hard real-time. Robotic systems are very dependent on the computational capabilities of the controlling computer, since they use computationally intensive algorithms. The computational requirements in robotics grows faster than the advancements in computer power (Fijany and Bejczy 1992). In addition the current serial algorithms are near optimal and there is unlikely to be much improvement in these algorithms, so parallel algorithms may offer an alternative.

2.2.1 Parallel Classification

There are various ways to classify parallel machines. A diagram of the parallel architecture types is shown in Figure (2.2). The architecture is the physical layout of the system, it can be general or specially constructed to suit a particular purpose. The popular classifications are: (1) Coupling; (2) Granularity; and (3) Flynn's classification.

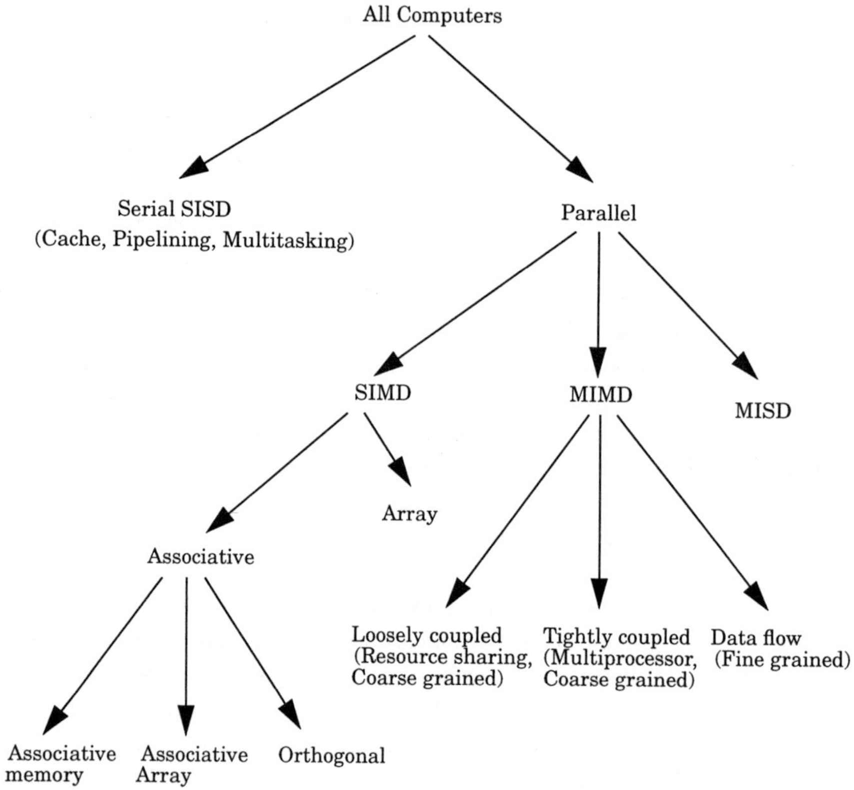

Figure 2.2: Computer architecture family tree

The coupling refers to the physical network topology and means of communication (Lawson 1992). The system can be: (1) Tightly coupled (multiprocessor system) where the processors share the same internal communication medium (e.g. a bus), operating system and memory; (2) Closely coupled (multicomputer system) where the processors share the same internal communication medium, but have separate operating systems and memory; and (3) Loosely coupled (multicomputer system) where each processor has a separate internal communication medium, operating system and memory.

Granularity refers to the level at which the parallelism is applied, i.e. the size of an individual task to be executed on a parallel processor (Stone 1990). These levels range from coarse to fine grained, and are: (1) Job level - between jobs and phases of jobs; (2) Program level - between parts of a program (procedures and loops); (3) Interinstructional level - between phases of an instruction cycle; and (4) Intrainstructional level - between bits in arithmetic circuits (Zomaya 1996).

Flynn (1966) classified the architecture of a computer in a macroscopic way on how the machine relates its instructions to the data being processed. The four classifications are based on whether there are single or multiple instruction or data streams. They are: (1) SISD; (2) SIMD; (3) MISD; and (4) MIMD.

2.2.1.1 SISD Machines

This is the most common computer architecture. There is a single processing unit that receives instructions from a single source (control unit) that operate on data from a single stream (memory), as shown in Figure (2.3). All von Neumann machines belong to this category and everything is performed sequentially. It cannot do any parallel processing.

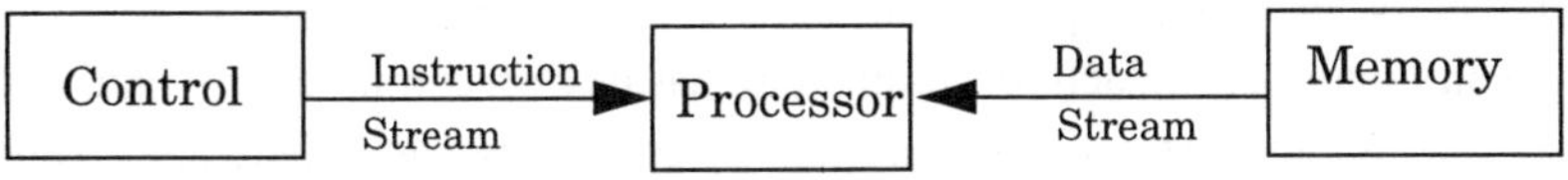

Figure 2.3: SISD architecture

2.2.1.2 SIMD Machines

In this type of machine, there are N identical processing units (Processing unit i; $i = 1, 2, ..., N$), as shown in Figure (2.4). Each processor contains its own local memory (LM-SIMD), where the communication takes place over an interconnection network, or shares memory (SM-SIMD), where the communication takes place within the common memory pool via buffers in memory (Moldovan 1993). There is a single control unit and instruction stream, and N data streams (Data stream i; $i = 1, 2, ..., N$). All operations are performed synchronously. Vector processors fall into this category. SIMD machines are extremely efficient in handling matrix and vector operations where there is inherent parallelism in the data.

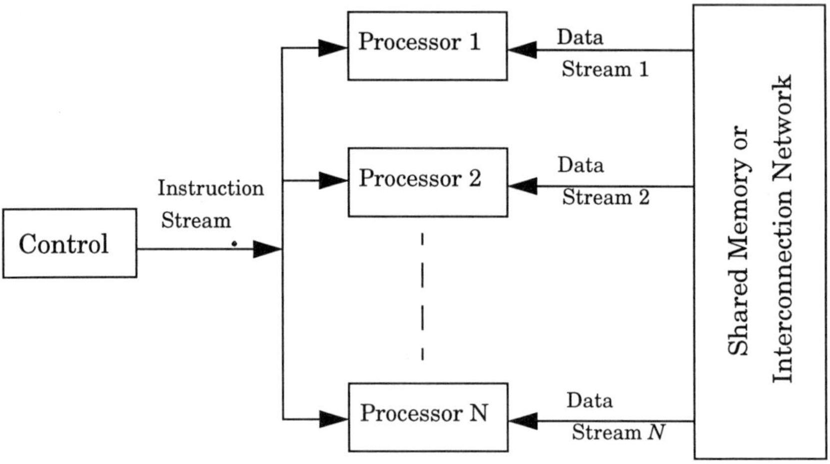

Figure 2.4: SIMD architecture

2.2.1.3 MISD Machines

In this class, there are *N* processing units (Processing unit *i*; *i* = 1, 2, ..., *N*), as shown in Figure (2.5). Each processing unit has its own control unit (Control unit *i*; *i* = 1, 2, ..., *N*), but share a common memory containing data. There are *N* separate instructions (Instruction stream *i*; *i* = 1, 2, ..., *N*) that operate simultaneously on the same item of data. Each processing unit does different things to the same data. This type of architecture is very rare and impractical. Systolic arrays fall into this category (Hwang 1993).

2.2.1.4 MIMD Machines

These machines are the most general and most powerful (Akl 1989). In this machine there are *N* processing units (Processing unit *i*; *i* = 1, 2, ..., *N*), along with their own instruction streams (Instruction stream *i*; *i* = 1, 2, ..., *N*) from their own control units (Control unit *i*; *i* = 1, 2, ..., *N*). Each processing unit receives data from its own data stream (Data stream *i*; *i* = 1, 2, ..., *N*), as shown in Figure (2.6). This machine is like a collection of SISD machines operating together asynchronously.

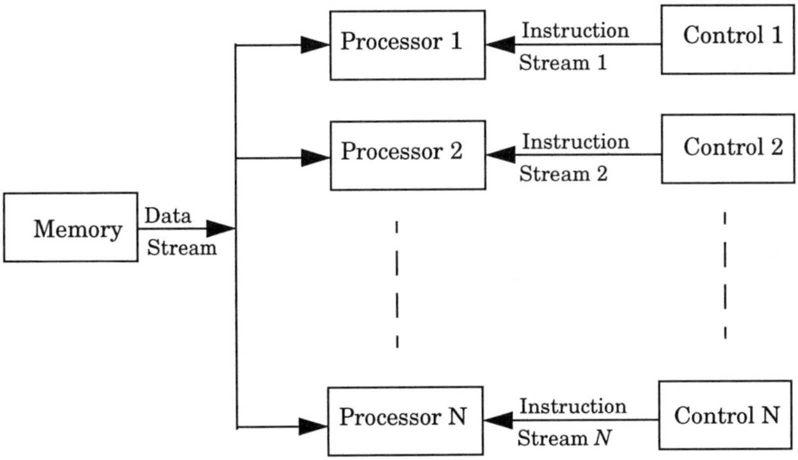

Figure 2.5: MISD architecture

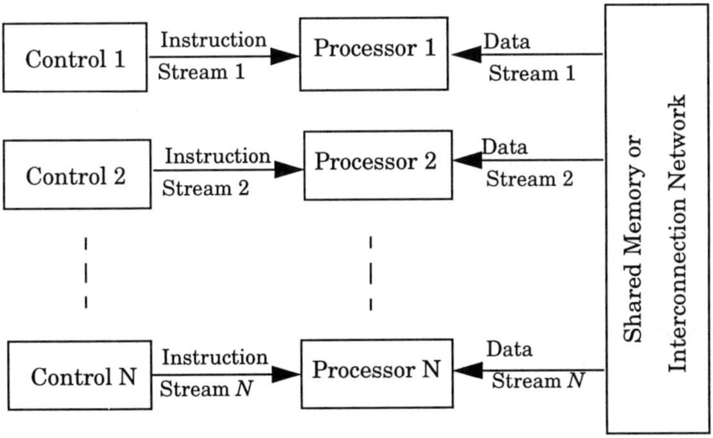

Figure 2.6: MIMD architecture

There are several varieties of MIMD machines. These range from fine-grained to coarse grained, and the coarse grained systems can be further subdivided into loosely coupled and tightly coupled systems. Data flow machines are fine grained systems, and are data driven systems (Dennis 1980). Unlike von Neumann machines, instructions are activated by the availability of their operands and not under control of a control unit. In the loosely coupled coarse grained systems each processor contains its own local memory and is connected to the others via an interconnection network and shares various resources on the network. The main drawback is that it is difficult to increase the number of processors on the system since this may increase the memory access contention depending on the network used (due to two or more processors attempting to access the memory simultaneously). But, it is well suited to handle tasks with high levels of interaction. In the tightly coupled coarse grained system each processor shares memory from a central memory pool. They are best suited to tasks where each processor works independently with minimal interaction, since the inter-processor communication efficiency is dependent upon the network. SPMD machines are a subclass of MIMD machines - each processor runs the same program asynchronously.

2.2.2 Network Topologies

The interconnection network is the device used to connect processors (and memories - in the case of shared memory systems) together. The structure of the network, the processor type, and the way in which a task is implemented on the system will determine its performance. Figure (2.7) shows the main network topology types. Common topologies include: bus architectures, ring networks, crossbars, meshes, shuffle exchanges networks, and hypercube ensembles (Seigel 1990).

In the shared memory architecture, the processors communicate through the shared memory. While in the distributed memory architecture, all communication between processors is performed by message passing. The distributed memory architecture may be either static or dynamic. In a static network all the communication links are fixed, unlike a dynamic network where the link connections can change while a program is running. A network is said to be fully connected if each processor is connected to all the other processors in the system.

The main factors affecting the choice of interconnection network for a particular application or set of applications are:

- Communication distance - The distance a message must travel between the sending and the receiving processors. It should ideally be as short as possible.

- Expandability - The ease at which the network can be extended to include more processors.

- Fault tolerance - The ability to handle failure of one or more processors. This is usually accomplished by using redundant networks so a message can travel along another path should a part of the network fail (Siegel 1990). The redundancy can also relieve network traffic contention by allowing a greater choice of communication paths.

- Bandwidth - The rate at which data can be sent over the network.

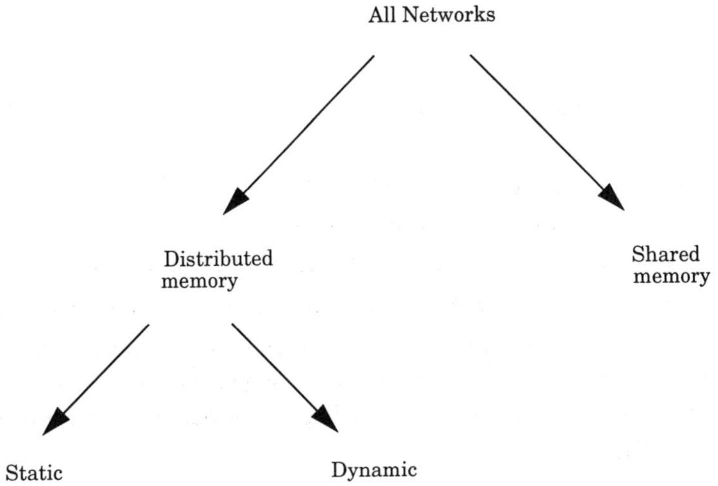

Figure 2.7: Network topology family tree

2.2.3 Performance Measures

The main objective of parallel processing is to speed-up the processing of a task. Performance degradation usually occurs as a result of either: (1) The number of tasks being greater than the number of processors and therefore a less efficient algorithm will be used; (2) Precedence relationships between the tasks requiring them to run in a specific order; and (3) Communication overheads. Optimum performance is achieved when there is maximum parallelisation and minimum interprocess communication. There are various ways to measure the performance improvement, to rate how well the parallel machine operates. The performance of computers is often reported as the number of clock cycles per second, the MIPS/ MFLOPS rating, or its benchmark rating from a particular benchmark kernel. The most obvious measure is the running time of an operation, which is the time elapsed from the time the operation starts until the moment it terminates (Akl 1989). Other common measures are given below:

2.2.3.1 Speed-Up

$$S = \frac{T(1)}{T(N)} \qquad (2.1)$$

or

$$S^* = \frac{T^*}{T(N)} \qquad (2.2)$$

The speed-up (2.1) indicates the quality of a parallel algorithm. $T(i)$ is the time taken on i processors, T^* is the execution time of the best available sequential algorithm, S^* is the speed-up in this case (2.2), and $S^* \le S$. Ideally, a speed-up of N for N processors would give the maximum speed-up. But this is not always attainable in practice, since not all tasks can be decomposed into N subtasks requiring $(1/N)^{th}$ of the time taken by one processor to solve the original problem. Also the structure of the parallel processing machine may impose restrictions on the running time (Zomaya 1996).

2.2.3.2 Maximum Speed-Up

$$S_N \le \frac{1}{f + (1-f)/N} \le \frac{1}{f} \qquad (2.3)$$

The speed-up is limited by the amount of inherent parallelism in the algorithm. f is the fraction of computation that must be done sequentially in an N processor system.

2.2.3.3 Communication Penalty

$$cp_i = \frac{E_i}{C_i} \tag{2.4}$$

The communication penalty is effect of the interprocess communication on processor *i*. Where E_i is the total execution time spent by processor *i* to run the algorithm, and C_i is the corresponding time for communications. Communication will occupy some of the processors time and other resources. A low ratio usually indicates that there is a relatively high amount of communications which may be due to an inefficient algorithm. On the other hand, a high ratio may show that there exists a poor exploitation of parallelism.

2.2.3.4 Efficiency

$$\phi = \frac{S}{N} \tag{2.5}$$

The efficiency of the parallel algorithm is a measure of how well the parallel system has been utilized. Ideally f should be one - perfect utilization of processors - but it is less than one in practice.

2.2.3.5 Utilization Factor

$$U = \frac{O(N)}{NT(N)} \tag{2.6}$$

The utilization factor measures the efficiency of resource utilisation (Zomaya 1996). Where *O(N)* is the actual total number of unit operations performed by an *N*-processor machine, while *NT(N)* represents the number of operations that could have been performed with *N* processors in *T(N)* time units.

2.2.4 Parallel Algorithms

The value of a parallel algorithm can be determined by a performance measure (as shown in the previous section). The architecture of the parallel system has a direct influence on way the algorithm is implemented and even the type of algorithm used. The choice of the architecture and the amount of effort put into developing a suitable algorithm is often governed by economic factors. The mapping of the algorithm onto the particular network can either be: (1) Static, where the mapping is decided before the algorithm is executed; or (2) Dynamic, where the process allocation, and perhaps, even the layout of a flexible interconnection network, is determined while the parallel program is executing. The parallel algorithms can also be: (1) Synchronous - tightly coupled with identical data; (2) Loosely synchronous - tightly coupled without identical data; (3) Asynchronous - little or no coupling between the processes; and (4) Embarrassingly parallel - independent processes (Mattson 1996).

Two common categories of parallel programs are: (1) Divide-and-Conquer; and (2) Master/Slave. Other categories can be found in Zomaya (1996).

The **divide-and-conquer** algorithm branches out from a root process, and the spawned processes can further subdivide iteratively. Each subprocess works on one part of the task and as soon as it has completed its function it returns the result to its source. The results progressively work their way back to the root until the task has finished. The speed-up of this type of algorithm is limited to (Lewis and El-Rewini, 1992):

$$\frac{N}{\log N} \tag{2.7}$$

The **master/slave** approach divides the task into separate components. These components are executed by the slave processors. The master processor supervises and coordinates the whole operation.

2.2.5 Real-Time Systems

An important application of parallel processing is in real-time systems. A real-time system is *a system that assures that controlled activities "progress" and that stability is maintained and further, that the values of outputs and the time at which the outputs are produced are important to the proper functioning of the system* (Lawson 1992). Recently there has been significant growth in real-time systems. This growth is related to the growth in computing power. Real-time systems are generally used to process vast amounts of information quickly or handle a large number of calculations quickly.

For a real-time system to successfully work it must operate correctly and produce the required results on time. It must also run continuously in a reliable and predictable manner. Real-time systems can either be soft or hard. In soft real-time systems a catastrophe will not result if the system is unable to comply with its requirements occasionally. But a catastrophe will result in a hard real-time system if it fails to deliver an appropriate result on time.

Real-time processes can be classified based on their periodicity, importance and whether they are static or dynamic (Lawson 1992; Ramamritham *et al.* 1989):

- Periodic (deterministic) processes - These are executed on a regular basis.
- Aperiodic (non-deterministic) process - These are executed when required, such as a particular event occurring.
- Critical processes - These must meet their required deadlines, otherwise the result could be disastrous.
- Essential processes - These should meet their deadlines for proper function of the system. But failure will not cause a disaster.
- Non-essential processes - These may miss their deadlines with no short term consequences to the system.
- Static processes - These processes are a permanent part of the system, and are therefore inflexible.
- Dynamic processes - These processes are created and removed, as required, during the operation of the system.

The most important aspect of a real time system is that it meets its deadlines, otherwise the result may be useless. The following factors are also significant (Lawson 1992):

- Predictability - The behavior of the system (especially the time required) must be deterministic.

- Supervision - The system must be able to be monitored to ensure correct operation.

- Synchronization and communication - The processes in the system must have their function synchronized to ensure the correct sequence of certain operations is enforced - some operations can only be performed in a strict sequence.

- Deadlock avoidance - The system must be able to respond when two or more processes are blocked to enable the system to proceed. Deadlock is *the state in which two or more processes are deferred indefinitely because they are mutually holding resources for each others progress* (Lawson 1992).

2.3 Master/Slave MIMD Implementation

There may be several ways to implement a sequential algorithm in parallel on an MIMD system. In cases where several processes can run in parallel, each performing the same operation on different data, with comparatively little communication, a master/slave approach can be used. The main program runs on the master processor and controls the operation. The slaves run on each of the other processors asynchronously. The master sends data to the slaves and after they have finished, they return the result back to the master and wait for more data. Each processor has its own memory and communication takes place via message passing.

The pipeline as shown in Figure (2.8) has been used in this work. The pipeline is a common network type and it is easy to transfer data through it with a simple multiplexer process on each node. Here all messages are passed along a pipeline from node to node until they reach their destination. The multiplexers control the communication through the slave processors (the slave processes and multiplexers run together simultaneously on multitasking processors). If there is a lot of data going through the system at any instance then delays may occur in the multiplexer processes. The multiplexers will also cause delays for the slave processes in processing the data as they will also be using some of the processors time and resources.

The fault tolerance of a pipeline is dependent upon which processor fails. If a slave processor fails on a pipeline network then all the subsequent processors will become unreachable.

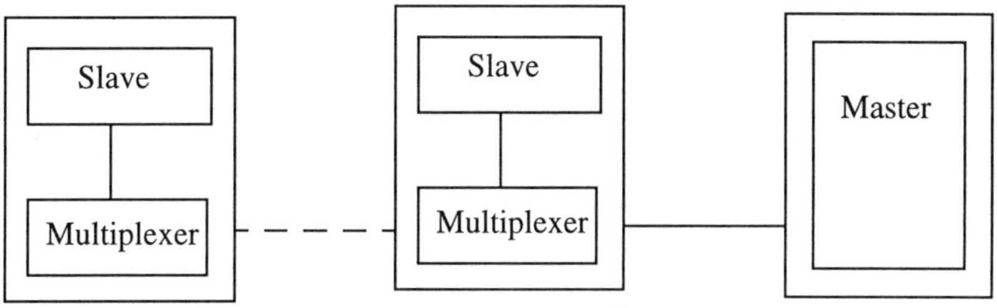

Figure 2.8: Pipeline network

In this work each slave is allocated an identification number (1, 2, ..., N) and the master is allocated number zero. Data is transferred between the master and the slaves in packets of the form in Figure (2.9). All the data that needs to be transferred is packed into a buffer and sent with the header attached.

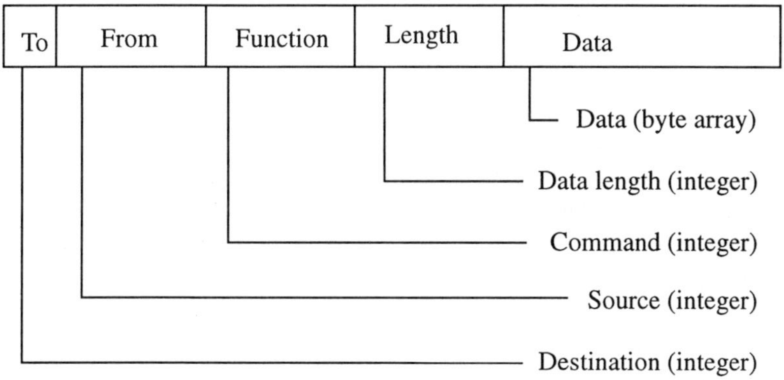

Figure 2.9: Data packets

The header consists of the following parameters:

To (integer): Intended destination (for the multiplexer process),

From (integer): Source of the packet (for the multiplexer process),
Function (integer): Purpose of the packet (for the slave process),
Length (integer): Length of the data in the packet (for both the slave and multiplexer processes), and
Data (byte array): Actual data (for the slave process).

If there are more items of data that need to be processed in parallel than there are slave processors, then there are two possible solutions. Either have more slave processes running on each slave processor (if it supports multitasking), or use a queuing system to transfer data between the master and the slaves when appropriate.

In a queuing system, if there are *N* slave processors in the system and *M* items of data that need to be processed by the slaves ($M > N$), then the first *N* items of data are sent to slaves (one for each). When a slave has finished processing the data the result is returned to the master. The master can then send the next item of data in the queue to the available slave. This process then continues until all the data has been processed. The main drawback with this approach is that there is a delay from when a slave returns results until it receives new data. Two possible ways exist to eliminate this latency. The first is to distribute all the data to the slaves (based on their individual performance and communication overheads - i.e. if there is an uneven distribution of data then the excess should be placed on the closest processors (assuming all processors have equal abilities) to reduce the final slave to master communication time). The second is to use multitasking (if it is available) and distribute *M* slave processes on the *N* slaves processors. An outline using C-like syntax of the queuing operation is shown in Figure (2.10).

The operation of the multiplexer process is straightforward. Two parallel processes are spawned in parallel with the slave process, one handles packets travelling from the master and the other handles packets travelling towards the master, as shown in Figure (2.11). The incoming switch handles packets coming in from the master. If a packet is destined for that particular slave then it is sent to the slave process, otherwise it is passed onto the next slave further down the chain. The outgoing switch handles packets travelling towards the master. Any packets received from the slave process and other slaves further down the chain are sent towards the master. There is no need for a multiplexer at the end of the chain (assuming the "to" header component is always correct in each packet). The two switches run together in parallel to avoid deadlock occurring when two slaves are trying to simultaneously transmit packets to each other. An outline using C-like syntax of the multiplexer operation is shown in Figure (2.12).

```
for (i = 0; i < number_of_processors; i++)
        send_data(i,data[i]);              /* Send first data out */
in_counter = 0;                            /* Amount of data received */
out_counter = number_of_processors;        /* Amount of data sent */
while (in_counter < amount_of_data) {
        free_processor = recieve_data(result[in_counter]);
                                           /* Receive data */
        in_counter++;
        if (out_counter < amount_of_data) {
               send_data(free_processor,data[out_counter]);
                                           /* Send more data */
               out_counter++;
        }
}

Array: data[i]: Array of output data, i = data item number,
               0 <= i < amount_of_data.
Array: result[i]: Array of input data, i = data item number,
               0 <= i < amount_of_data.
Function: send_data(i,data[i]): Send data element i to processor i.
Function: recieve_data(result[i]): Get data from slave when ready and return
               identity of slave.
```

Figure 2.10: Queuing pseudo-code

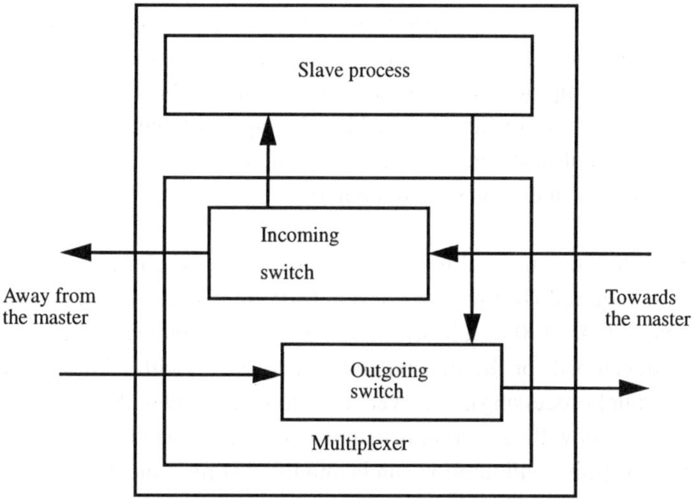

Figure 2.11: Multiplexer operation

```
/* Incoming process */

while (0 == 0) {
        destination = get_data_from_master(buffer);
        if (destination == slave_number)
                send_data_to_slave(buffer);
        else
                send_data_to_next_slave(buffer);
}

/* Outgoing process */

while (0 == 0) {
        source = next_ready_source();
        if (source == slave)
                get_data_from_slave(buffer);
        if (source == next_slave)
                get_data_from_next_slave(buffer);
        send_data_to_master(buffer);
}
```

Figure 2.12: Multiplexer pseudo-code

2.4 Transputers

The transputer is a processor specifically designed to support parallel processing. The T805 transputer (Figure 2.13) is a 32 bit CMOS architecture (INMOS 1992), with a 64 bit on-chip floating point unit, 4K on-chip RAM and four 5/10/20 Mbits/sec serial links. They are capable of addressing up to 4 Gbytes of external memory and to support multitasking. The transputers are an MIMD architecture using distributed memory rather than shared memory.

The four bi-directional serial links are used to connect each transputer to others or to a host if required (In this work seven T805 transputers are used with an IBM compatible 486DX-100 as a host). They can be programmed in Occam (a parallel language developed for the transputers) and ANSI C (with additional libraries to support parallel processing), and even a mixture of the two. A configuration file is used to specify how the program is mapped onto the network and how the network is constructed (this configuration can be modified at run time under Occam but not C).

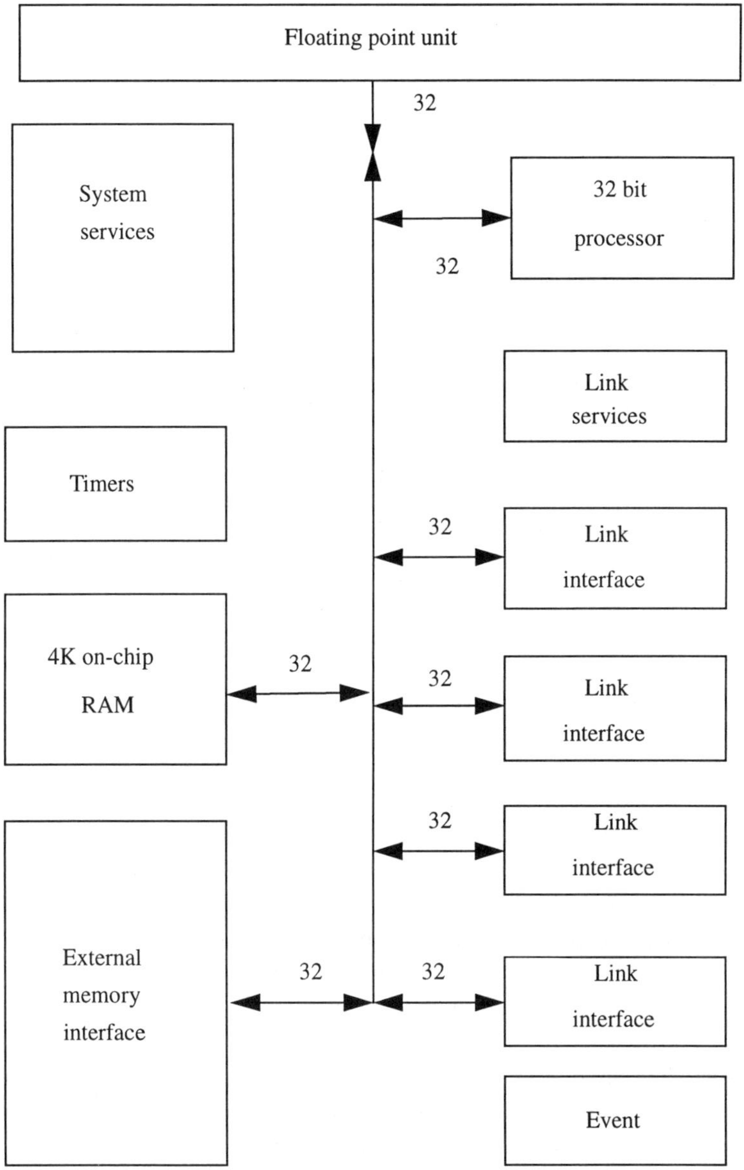

Figure 2.13: T805 architecture

The transputers are mounted on a B008 TRAM motherboard (Figure 2.14). The motherboard contains slots for transputers, memory, the C004 crossbar switch, and a C012 host interface. The TRAM can hold up to ten transputers. Some of the link connections are fixed (hardwired), and the rest can be configured (softwired) through the crossbar switch. Each slot can hold any transputer.

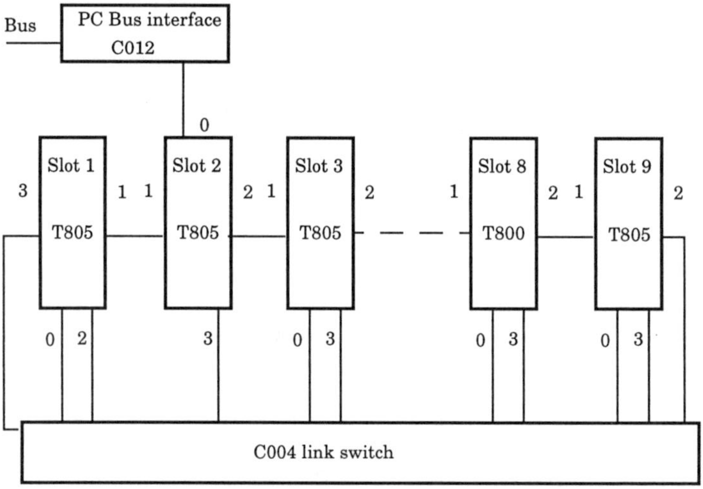

Figure 2.14: TRAM motherboard

The PC bus interface connects the PC bus to the TRAM. The bus interface converts data travelling between the host and the TRAM. Data travelling between the host always goes through link 0 of slot 0. The link switch provides a full crossbar switch between 32 link inputs and 32 link outputs. The network configuration processor (slot 1) issues instructions to the link switch to establish the link connections. Data goes to it through link 3 from link 1 of slot 0.

The master/slave approach is implemented on a transputer network with the master placed on the first processor (which is attached to the host). Each slave is placed on subsequent processors that are chain linked together from the master. On each slave processor a multiplexer process was also implemented to handle communications (as this is not automatically handled on transputers).

2.5 Summary

This chapter has presented an introduction to how parallel processing is used in this work. Parallel machines are often used in real-time systems to enable them to achieve their goals. The main types of classification of parallel systems have been discussed, along with various issues affecting the design and performance of parallel systems. The two types of systems that have been employed are explained, and how a master/slave system can be used on them. Some of the work to follow in later sections will use parallel processing to speed-up calculations.

CHAPTER 3

PATH PLANNING

3.1 Introduction

Robot manipulator path planning is the process of creating collision free paths for robot manipulators to enable them to carry out a desired task. Path planners remove the need for an operator to explicitly specify a path. They enable the operator to concentrate on the task, rather than determining how the manipulator should move. A path planner only requires a starting point and goal to be initially specified. The path planner is then expected to be able to generate a suitable trajectory for the manipulator to follow.

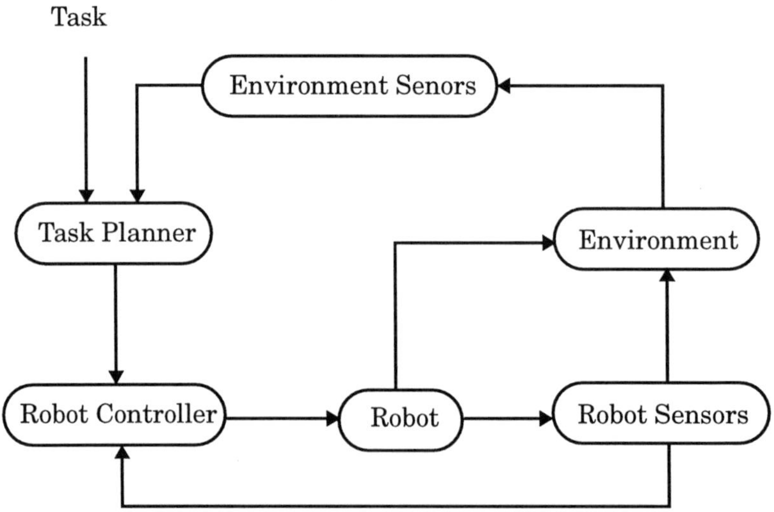

Figure 3.1: Robot control system

The robot control system (Figure 3.1) is composed of several major components. It generates the required control signals for the actuators (mechanism to drive the joints of the manipulator) and enables it to carry out a particular task, using information about the environment.

The task planner converts a high level language task-specification to a robot-level specification that can be interpreted by the control unit (Jacak 1990). The objectives to be attained are specified by the user, rather than the means to achieve the objectives. The task planner also requires information about the robot itself, the environment and obstacles in the environment it is working in. Information about the environment is provided by sensors.

The operation of the task planner is depicted in Figure (3.2). It must go through several stages to produce a suitable trajectory. Initially complex tasks (requests from the operator) are decomposed into elementary operations (a series of necessary incremental actions). Each operation is represented by its initial and final goal configurations. This high level interface is known as the operations planner.

The next stage is the path planner. It takes the initial and goal configurations from the operations planner and generates a collision free path through the workspace. The path itself is a series of points from the start to the goal. It should be short and avoid any unnecessary movements. The collision-free path is then used by the trajectory planner to create a continuous-time joint path (trajectory) for the manipulator.

Trajectory planners can generate the following kinds of trajectories (Critchlow 1985):

- Sequential joint control - Each joint is moved separately in a specified sequence. No two joints move simultaneously. A manipulator using this type of motion moves relatively slowly.

- Slewing (uncoordinated) joint control - Each joint is moved simultaneously with the same rotational velocity. The path followed by the manipulator is unpredictable due to the complex interaction between the links.

- Interpolated joint control - All the joints move together in a unified fashion. They start and stop at the same time. The intermediate path is more constrained than in the previous two methods, but the acceleration and velocity of each joint are variable to achieve the coordinated motion.

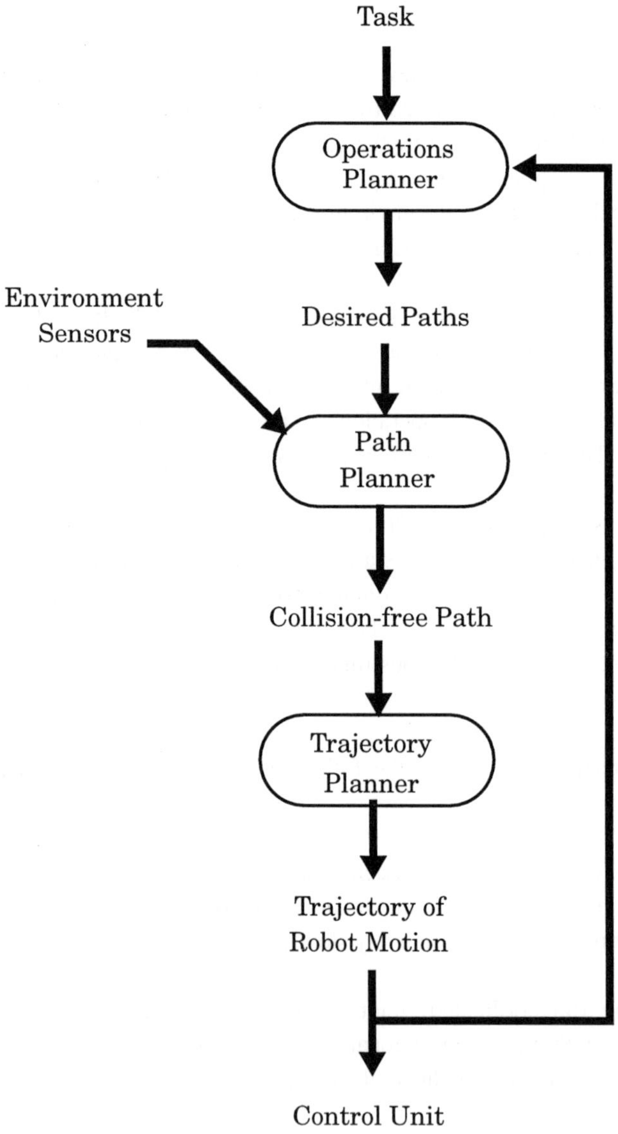

Figure 3.2: Task planner

- Straight-line control - The joints are moved in a particular manner to ensure that the end-effector follows a set path in its workspace. The manipulators speed is limited by the slowest joint. The path may be simple in cartesian space but complex in joint space.

The resultant joint trajectory generated by the task planner is used by the controller. The controller uses feedback about the state of the manipulator and the input from the task planner to produce the required control signals for the manipulator to maintain the correct trajectory. The state of the manipulator is the position and velocity of each joint along with the forces and torques applied by the end-effector. The control signals generated are used to control the actuators for each link. Each actuator must produce a specific force or torque to enable the manipulator to manoeuver correctly.

This chapter will provide an overview of the path planning problem. Initially terms will be defined (Section 3.2). Then, an overview of the motion planning problem (Section 3.3), motion planning algorithms (Section 3.4), and general approaches to solving the path planning problem (Section 3.5) will be presented. Finally, the various methods used in path planning and other work in the area will be reviewed (Section 3.6). For more comprehensive information about the structure of manipulators refer to Chapter (5).

3.2 Defining Terms

Cartesian space (task, world or operational space): *The three dimensional physical space. The robots and obstacles exist in this space. Coordinates are generally specified in terms of (x,y,z). It is easy to specify the desired task and position and orientation of the end-effector in cartesian space. The dimension of cartesian space is always fixed which places an upper bound on the complexity due to the dimensionality. The problem with the cartesian space representation is that it cannot be used to directly control the manipulator. The joint variables must be derived using an inverse kinematics procedure and this has problems with degeneracy and singularities.*

Fine-motion planning: *Planning the path of the manipulator to enable it to come into contact and interact with objects in the workspace. It is used when grasping and moving objects. Constant feedback from sensors is required to enable the correct level of force to be applied.*

Gross motion planning: *Planning the path of a manipulator so that it does not come into contact with any obstacles in its workspace. The obstacle avoidance is the primary goal of this type of planning, other factors such as optimizing the path are secondary.*

Manipulator: *A mechanical robot arm consisting of a series of links and joints.*

Workspace: *The subset of cartesian space that is composed of all the points that the end-effector of the manipulator can reach without violating any constraints imposed on the manipulator due to its structure. There are assumed to be no obstacles in the workspace.*

Workspace envelope: *The surface of the workspace, i.e. the maximum limit on where the manipulator can reach.*

Reachable-space: *The subset of the workspace that the end-effector (robot hand) can reach without coming into contact with obstacles in the workspace (Refer to Figure 3.3).*

Obstacle: *Any object in the environment that can restrict the motion of a manipulator (Refer to Figure 3.3).*

Safety margin: *The area around an obstacle that a manipulator should avoid in order to operate safely. It takes into account the positioning accuracy of the manipulator (Refer to Figure 3.3).*

Pseudo-obstacle: *The new representation of an obstacle with the safety margin taken into account, i.e. Obstacle + Safety margin.*

Obstacle shadow: *The region in the workspace that cannot be reached by the manipulator due to the presence of an obstacle in the workspace. It also includes the pseudo-obstacle that created the shadow (Refer to Figure 3.3).*

Free space: *The areas in space where there are no obstacles.*

Configuration (joint) space: *The set of all configurations of a manipulator. The configuration of a manipulator is given in terms of joint variables - one for each joint. The dimension of the configuration space is the same as the number of joints*

(DOF). It can, therefore, become very complex in manipulators with a high number of joints. Obstacles can be represented in the configuration space as sets of unreachable points. To find this set of points a mapping process must be used. The manipulator maps into a single point - every point in the configuration space represents a particular configuration of the manipulator.

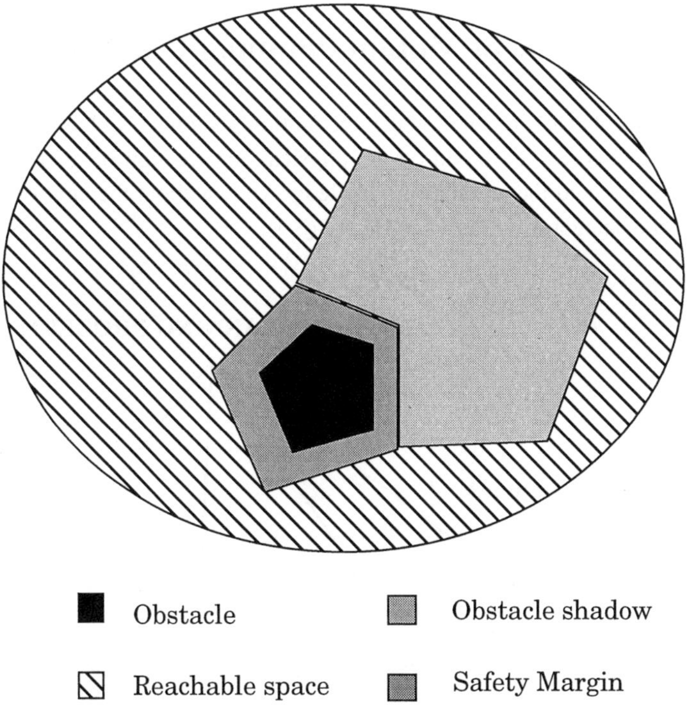

Figure 3.3: Environment

Path: *The curve in the configuration space representing the motion of the manipulator. This is either a mathematical expression or a sequence of points. It does not consider time or velocity, only the joint angles.*

Trajectory: *The path with time included. Each point along the path represents the position of the manipulator at a particular time. How the manipulator follows the path depends on its mechanical structure.*

Feasible path: *A collision free path.*

Generalized mover's problem: *Locating a collision-free path for a manipulator (Hwang and Ahuja 1992).*

3.3 Classification of Motion Planning Problems

There are several categories into which the motion planning problem can be classified. These categories are based on the physical construction of the environment and manipulator and the amount of knowledge known about the environment. An outline of the categories is shown in Figure (3.4).

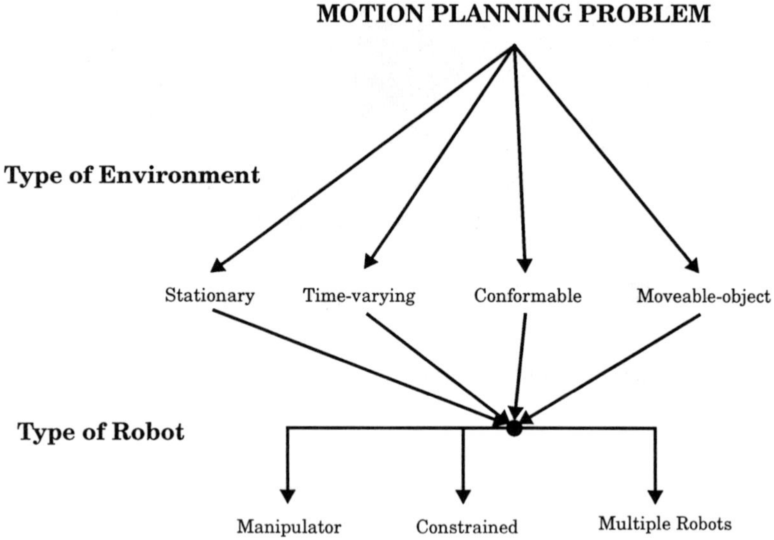

Figure 3.4: Taxonomy of motion planning problems

There are different types of environments for a manipulator to operate in. In a **sta-**

tionary environment all the objects are stationary and their shape is fixed. If the objects in the environment are moving, then the problem becomes a **time-varying** problem, otherwise it is **time-invariant**. Where the objects can change shape the problem becomes a **conformable** one, otherwise the problem is **nonconformable**. If the manipulators move any of the objects then the problem becomes the **movable-object** problem.

If there is more than one manipulator to which to plan the path and the manipulators have overlapping workspaces then the problem is a **multimovers** problem. If there are any restrictions on the manipulator (apart from those placed on it by objects in the workspace) the motion planning is **constrained**, otherwise it is **unconstrained**. The constraints can be velocity and acceleration bounds or physical limitations on how the joints move.

The amount of information available to the motion planner about the workspace may be complete - where everything about all possible obstacles is known in advance - a **static** problem. If not all information about the workspace is known then the problem is a **dynamic** one. In the dynamic case, information gathered about the environment is stored, but it must be updated if something changes. Most of the motion planning problems have used the static approach (Hwang and Ahuja 1992).

3.4 Classification of Motion Planning Algorithms

As with the classification of motion planning problems, the algorithms used to solve them are classified. The classification is based the nature of the algorithm used and how it works. Figure (3.5) depicts the motion planning algorithm categories.

The **completeness** of the algorithm refers to its ability to locate a solution to the problem. If the algorithm is **exact**, then it is guaranteed to find a solution or to show that no solution exists. Exact algorithms are computationally expensive (Hwang and Ahuja 1992). A **resolution complete** algorithm is exact for a given resolution of the workspace. Most algorithms use discrete representations of the workspace (Paden *et al.* 1989; Kondo 1991). The accuracy can be increased by increasing the resolution of the workspace. As the resolution increases the discretization approaches a continuum, and the algorithm will become exact. On the other hand, if the probability of an algorithm finding solution is close to unity then it is a

probabilistic complete algorithm. A **heuristic algorithm** is not guaranteed to find a solution, even if one exists, or the solution found may not be the best. Heuristics are used to find a solution within a given error margin and usually work much faster than the previous three methods.

Global algorithms consider all the information in the environment and plan the motion accordingly. They are computationally expensive and the path planning must be performed off-line. **Local** algorithms only have a limited knowledge of their environment, but can be used on-line. The knowledge they have is that of their immediate vicinity. They can be used in conjunction with global methods to counter any unexpected changes that may affect the manipulator.

The operational space of the algorithm can be either **cartesian** space or **configuration** space. If the algorithm can be solved in polynomial time (i.e in P) then it is said to be **tractable**, otherwise if there exists an exponential time bound (i.e. it is in NP) then it is **intractable** (Latombe 1991). Chapter (4) will further discuss NP-completeness.

3.5 Path Planning

In general, the main steps to be considered when developing a path planning algorithm are (Hwang and Ahuja 1992):

- Determine the configuration parameters of the manipulator to develop the configuration space representation if configuration space is to be used.
- Develop a representation of the manipulator and objects.
- Select an appropriate motion planning method to solve the problem.
- Choose a suitable search technique to find a solution path and then locally optimize this path to shorten and smooth it.

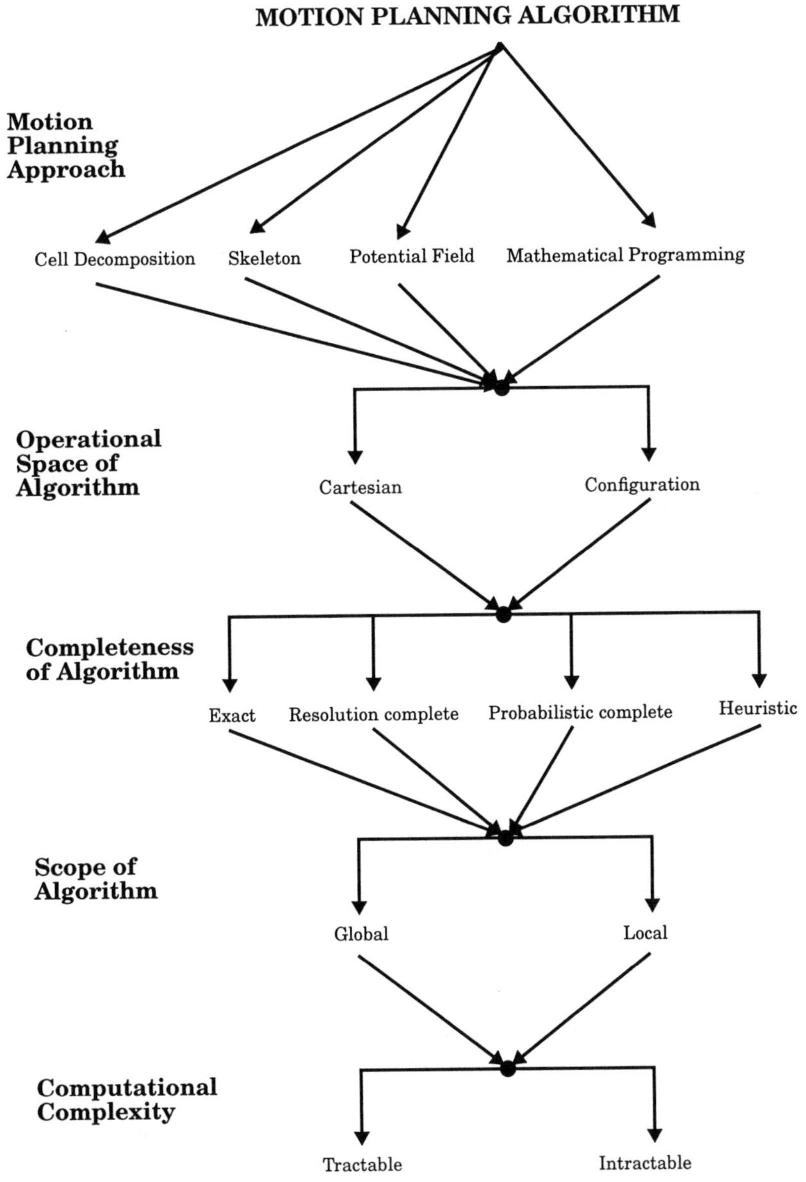

Figure 3.5: Taxonomy of motion planning algorithms

3.5.1 Configuration Space Representation

The notion of configuration space was developed by Lozano-Perez and Wesley (1979). Prior to this, all path planning methods depended on the use of cartesian space to represent manipulators and their environment. The configuration space approach reduces the problem to that of moving a point through n-dimensional space (n = manipulator DOF). Configurations that cause collisions with objects are known as configuration space obstacles.

There are seven main methods used to calculate the configuration obstacles, all of which can be applied to any robot and environment (Hwang and Ahuja 1992), are:

- Point evaluation - this is the simplest and least efficient method. Each configuration of the manipulator is considered in turn and if at the configuration an intersection occurs then this configuration is part of a configuration obstacle.

- Minkowski set difference - the union of the Minkowski set difference between the areas occupied by the obstacles and manipulator forms the configuration obstacle. This method is used in two-dimensional problems (Latombe 1991; Zhu and Latombe 1991).

- Boundary equations - in this method a set of equations (constraints) are derived in terms of the joint variables that bring the manipulator into contact with an object. Evaluate the equations to determine if a configuration is feasible. The higher the DOF, the more complex this approach becomes.

- Needle method - a representation of the configuration space obstacles is generated by fixing all but one of the joint variables and then the allowable values of this free variable are computed using the boundary equations (Lozano-Perez 1983; Lozano-Perez 1987).

- Sweep volume method - in this method the most proximal link of the manipulator is moved through its entire joint range. The areas where no intersection occurs are recorded and in those areas the next link is moved through its entire range in the same way as the first link. This process continues through all the links until they all have been considered. This method is similar to the sequential search strategy (Gupta 1990).

- Template method - each object is represented by a set of "standard objects" for which the configuration space transformation is already known (Newman and Branicky 1990).

- Jacobian-based method - the Jacobian (a matrix that relates the velocities in configuration space to velocities in cartesian space) is used to locate free regions and occupied regions in configuration space.

The advantages of using a configuration space representation is that there are no complications with redundant DOF manipulators, multiple solutions, undefined states or singularities (Khatib 1986) and the path planning problem simply becomes a problem of moving a point through a n-dimensional space (Schilling 1990), where n is the DOF of the manipulator. The main problems with configuration space are the difficulty in mapping cartesian space into it, and its dimensionality. The dimension of configuration space is the same as the DOF of the manipulator. Figure (3.6) shows a two DOF planar manipulator, with both joints revolute and link lengths of one unit, and a square obstacle (the sides have a length of one unit). The configuration space representation, in terms of the joint angles, is shown in Figure (3.7), which, even for this simple case, was complex to derive.

3.5.2 Object Representation

Each object in the environment must be modelled and stored in a suitable data structure. Knowledge of the obstacles must be known before a collision free path can be obtained. In on-line planners, the knowledge is acquired from sensors as the manipulator moves through the environment.

The main methods used to represent the objects (refer to Figure 3.8) are (Hwang and Ahuja 1992):

- Grid array - the workspace is subdivided into a grid (Zhu and Latombe 1991; Kondo 1991; Latombe 1991). All cells have the same size. Each cell in the grid is marked if it is occupied by part of an object. The binary values one and zero are usually used to denote that a cell is occupied or unoccupied, respectively. The cells can also be marked according to the fraction of the cell that is occupied by the obstacle. The accuracy depends on the resolution and it requires a large amount of memory. It works well with objects having an irregular shape.

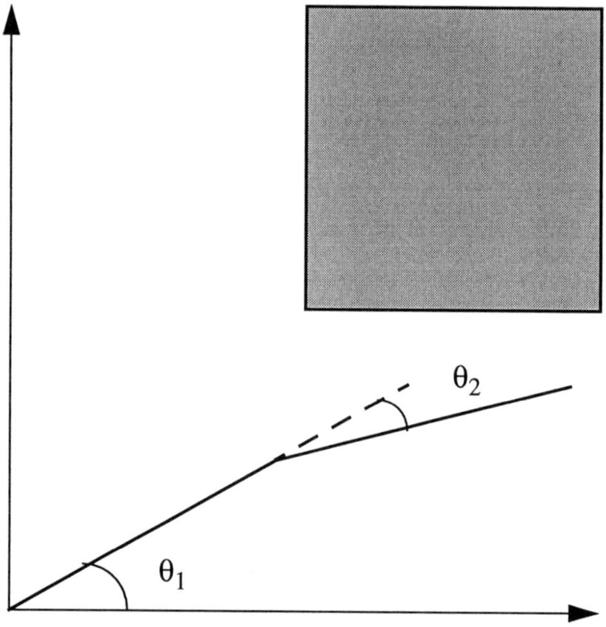

Figure 3.6: Simple planar manipulator

- Cell tree - variable size cells, rather than fixed sized cells, are used to represent the object in the same manner as the grid array. The workspace is divided into quarters (eighths in three dimensions). If any of the cells contain parts of an object, then those cells are further subdivided. This checking and subdivision continues until the resolution limit is reached. The cell tree is also known as a quadtree in two dimensions, or octree in three dimensions, and 2^n-tree in higher dimensions, where n is the dimension.

- Polygon representation - in this method each object is approximated by a polygon (Lozano-Perez 1983; Lozano-Perez 1987; Kim and Khosla 1992) or the union of several polygons (or polyhedra in three dimensions). This method uses relatively little memory (only the coordinates of the vertices are stored) and there exist known methods used to determine if polygons or polyhedra intersect.

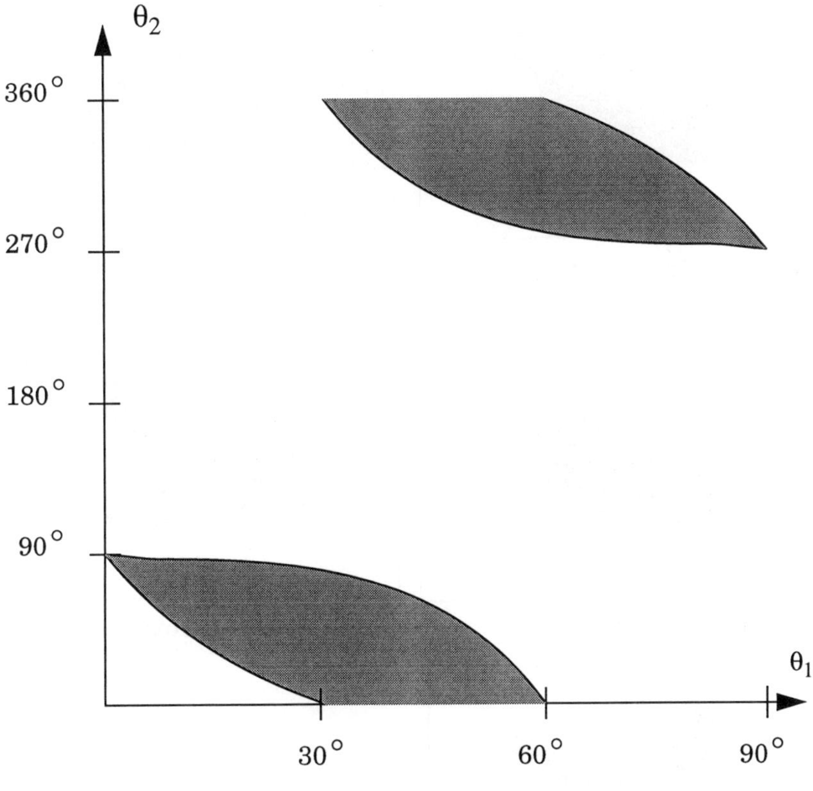

Figure 3.7: Configuration space representation

- CSG - the objects are represented as unions, intersections, and set differences of primitive shapes (Hwang and Ahuja 1992). Boundary representation is similar except that it uses the boundary features. Unlike previous methods, curved sections can be accurately represented.

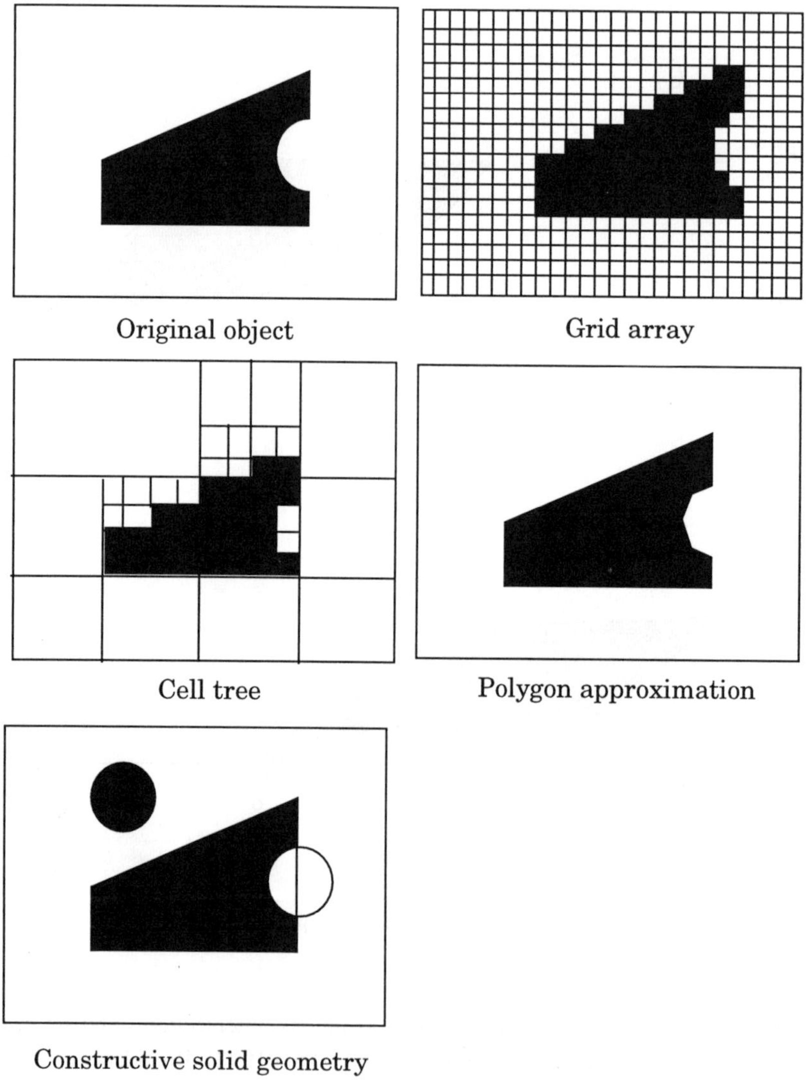

Figure 3.8: Object representation

3.5.3 Motion Planning Method

There are a large number of algorithms used to solve the motion planning problem. They vary in their scope - some can be used in general problems, while others apply only to specific situations - in some cases a combination of approaches can be used. The basic approaches used are (Latombe 1991; Hwang and Ahuja 1992):

- Skeleton (roadmap) methods - These methods reduce (or retract) the set of feasible motions onto a network of one dimensional curves. A feasible path between the initial and goal points is then found using a search technique. The initial and goal points do not often lie on any of the curves, so they are connected directly to the closest available point on a curve using another line. All possible paths must be represented, otherwise the method is incomplete, and the method requires a complete description of the workspace. The problem is further complicated when there are moving or shape-changing obstacles, and the problem has more than two dimensions. The advantage is that they will give the shortest path. The most common skeleton methods are: the visibility graph, Voronoi diagram, silhouette, subgoal and freeway net.

- Cell decomposition - In this technique the free space is decomposed into cells. A path is found from the starting point to the goal by moving from a cell to an adjacent cell. The decomposition can either be exact (exact cell decomposition - object dependent) or approximate (approximate cell decomposition - object independent). This approach will be discussed further in Chapter (7).

- Potential field - An artificial potential function is created which has a global minimum at the goal. Objects are represented by local maxima at their position. The manipulator starts at its initial point and is drawn to the goal by an attractive force and away from objects by repulsive forces. This approach was developed by (Khatib and Mampey 1978; Hogan 1985; Miyazaki and Arimoto 1984; Pavlov and Voronin 1984). This approach will be discussed further in Chapter (7).

- Mathematical programming - A set of inequalities is derived, representing the motion of the manipulator and the restrictions placed upon its motion by objects. The equations are then optimized to find a suitable path (Ge and McCarthy 1990). Numerical methods are often used due to the complexity of the non-linear equations.

3.5.4 Search Techniques

Once a suitable set of paths has been found, a search algorithm needs to be employed to locate solution. In off-line path planning all solutions can be considered, and therefore the optimal can be found. Not all search techniques used may find the optimal solution (e.g. heuristic search algorithms) but they may find a solution, within a particular set of requirements, quickly.

The most commonly employed methods are: (1) Depth-first search; (2) Breadth-first search; (3) Best-first search; (4) A* search; (5) Bidirectional search; and (6) Randomized search (such as simulated annealing and GAs). Hill climbing, random (not randomized) search, and enumerated searches are generally not used. There are numerous mathematical texts on these topics and a description of the search techniques used in this work will be given in Chapter (4).

After a path has been found, it is then optimized to shorten the path, smooth the motion and ensure adequate clearance is given around obstacles. Numerical optimization methods are often used.

3.6 Survey of Previous Work

Hwang and Ahuja (1992) provide a substantial survey of the gross motion planning problem. Tables (3.1), (3.2) and (3.3) summarizes the main approaches that have been used in the past for robot manipulators. Other methods not included are: (1) The classical mover's problem, which involves moving an object around other fixed objects; and (2) Point or circular robots. These other methods are not particularly relevant to this work. In Table (3.2) all algorithms are global. In Table (3.3) all algorithms are global and the robot is a point. The data presented in each column is defined as follows:

- The **approach** taken to solve the motion planning problem:
  - s = Skeleton,
  - cd = Cell decomposition,
  - pf = Potential field,
  - mp = Mathematical programming,

pp = Predefined priority (multimovers), and
npp = No predefined priority (multimovers).

- The author(s) of the **algorithm** with the year of publication.

- The **DOF** of the robot involved.

- The **scope** of the algorithm can be either global or local.

- The level of **completeness** is the extent that the algorithm has been tested, either a simulation or actual implementation.

- The **shape** of the manipulator involved.

- The shape of the **obstacle**s used.

- How **complete** the algorithm is:
  1 = An exact algorithm,
  1r = A resolution complete algorithm,
  1p = A probabilistic complete algorithm,
  2 = A heuristic algorithm that only fails when the environment is very cluttered, and
  3 = An algorithm that will often fail to find a solution unless the environment is relatively uncluttered.

- The **speed** at which the algorithm operates at:
  s = Under a minute,
  m = Between one and ten minutes,
  M = Ten to sixty minutes, and
  h = Greater than one hour

Approach	Algorithm	DOF	Scope	Completeness	Shape	Obstacles	Complete	Speed
s	Barraquand & Latombe 1990	arbitrary	global	simulation	polyhedron	arbitrary	1p	m
s	Chen & Hwang 1992	arbitrary	global	simulation	polyhedron	polyhedron	1r	m
s	Faverjon & Tournassound 1987	arbitrary	global	simulation	polyhedron	polyhedron	1p	M
s	Hopcroft *et al.* 1985	arbitrary	global	theoretical	line	circle	1	-
s	Lumensky 1986	3	global	simulation	line	polyhedron	1	s
s	Lumensky 1987	2	global	simulation	line	arbitrary (2D)	1	s
s	Lumensky & Sun 1987	2	global	simulation	line	arbitrary (3D)	1	s
cd	Brooks 1983	6	global	implementation	polyhedron	polyhedron	3	s
cd	Faverjon 1989	6	global	implementation	polyhedron	polyhedron	2	m
cd	Herman 1986	6	global	implementation	polyhedron	arbitrary	3	m
cd	Kondo 1991	6	global	simulation	polyhedron	polyhedron	1r	m
cd	Lozano-Perez 1987	6	global	implementation	polyhedron	polyhedron	1r	h

Table 3.1a: Summary of previous work on manipulators

Approach	Algorithm	DOF	Scope	Completeness	Shape	Obstacles	Complete	Speed
cd	Paden *et al.* 1989	6	global	simulation	arbitrary	arbitrary	1r	h
pf	Barraquand & Latombe 1990	arbitrary	global	simulation	polyhedron	arbitrary	1p	m
pf	Boissiere & Harrigan 1988	3	local	implementation	polygon	polygon	3	s
pf	Khatib 1985	arbitrary	local	implementation	polyhedron	polyhedron	3	s
pf	Khosla & Volpe 1988	arbitrary	local	simulation	polygon	polygon	3	s
pf	Newman & Hogan 1987	arbitrary	local	implementation	polygon	polygon	3	s
mp	Chen & Vidyasagar 1988	arbitrary	global	simulation	polygon	polygon	2	h
mp	Eltimsahy & Yang 1988	arbitrary	global	simulation	polyhedron	polyhedron	1	m
mp	Maciejewski & Klein 1985	arbitrary	global	simulation	polyhedron	polyhedron	2	m
mp	Shiller & Dubowsky 1988	arbitrary	global	simulation	polygon	polygon	2	h

Table 3.1b: Summary of previous work on manipulators

Approach	Algorithm	DOF	Robots	Implementation	Shape	Obstacles	Complete	Speed
pp	Erdmann & Lozano-Perez 1986	2	arbitrary	simulation	polygon or manipulator	polygon	2	m
pp	Freund & Hoyer 1988	2	arbitrary	implementation	point or manipulator	polygon	3	on-line
npp	Buckley 1989	2	arbitrary	implementation	square	none	2	s
npp	Chien *et al.* 1988	2	arbitrary	implementation	manipulator	polyhedron	3	s
npp	Fortune *et al.* 1986	2	2	simulation	manipulator	polyhedron	1	s/m
npp	Liu *et al.* 1989	2	2	simulation	circle	arbitrary	2	m
npp	O'Donnell & Lozano-Perez 1989	arbitrary	2	simulation	manipulator	polyhedron	2	m
npp	Schwartz & Sharir 1983	2	arbitrary	theoretical	circle	polyhedron	1	h
npp	Sharir & Sifrony 1988	2	2	theoretical	circle or manipulator	polygon	1	-
npp	Yeung & Bekey 1987	2	arbitrary	simulation	circle	polyhedron	2	s/m

Table 3.2: Summary of previous work on the multimovers problem

Approach	Algorithm	DOF	Implementation	Obstacles	Complete	Speed
s	Fujimura & Samet 1989	2	simulation	polygon	1r	m
s	Kant & Zucker 1988	2	simulation	polygon	2	s/m
s	Kehtarnavaz & Li 1988	2	simulation	point	3	s
s	Reif & Sharir 1985	2/3	theoretical	polytope	1	-
s	Canny & Reif 1987	2/3	theoretical	polytope	1	m/M
cd	Fujimura & Samet 1988	2	simulation	polygon	1r	m

Table 3.3: Summary of previous work in time-varying environments

There are many other examples on the various ways in which the collision avoidance problem has been solved. Cell decomposition has been used in this work and also in (Shiller and Dubowsky 1991). The most common approach has been to use potential fields (also used in this work), often in configuration space (Lozano-Perez 1986; Warren 1989; Warren 1990; Barraquand *et al.* 1992; Barraquand and Latombe 1991; Rimon and Koditeschek 1992). GAs, a heuristic search used in this work, have also been combined with potential fields or cell decomposition in configuration space (Shibata *et al.* 1992; Mazer *et al.*; Bessiere *et al.* 1993; Ahuactzin *et al.* 1993). Simulated annealing, another heuristic search, is an optimization algorithm that has also been used (Lee and Park 1991; Carriker *et al.* 1990; Janabi-Sharifi and Vinke 1993).

3.7 Summary

This chapter presented an overview of the motion planning problem to establish a foundation. In particular, defining what it is, and discussing the methods used to solve it. The motion planning problem is very complex and there are several stages in solving it. Numerous techniques have been applied in the past. This work will focus on the path planning and trajectory generation side of solving the problem.

CHAPTER 4

SEARCH TECHNIQUES

4.1 Introduction

Search and optimization techniques are used to locate optimal (or near optimal) solutions to problems. The techniques are either tailored to a specific problem, or are general algorithms applicable to various problems. The goal of optimization is to improve performance, but the way in which the improvement is attained is also significant. In complex systems the improvement process becomes significant. The goal of optimization may not be to attain the perfect solution, but rather to find a reasonable solution quickly.

The complexity or hardness of the problem is determined by the running time of an algorithm used to solve it. A problem is said to be in class P (polynomial) if it runs in polynomial time (it is tractable), i.e. if the worst case computation time, for n inputs, is $O(n^k)$ for some constant k (Cormen *et al.* 1990). The P class of problems is a subset of the NP (non-deterministic polynomial) class of problems. A problem is in NP if there is a polynomial time algorithm to verify the correctness of a solution to it. An NP-complete problem is one which is in NP, and is at least as difficult to solve as any other NP problem. The only known algorithms to solve NP-complete problems run in exponential-time. The question of whether P=NP is an open question since if a polynomial-time solution was found to any NP-complete problem then a polynomial-time solution could be found for all the other NP-complete problems. NP-complete problems form another subset of NP problems and are in some sense the most difficult ones to solve. The relationship between the classes is depicted in Figure (4.1).

The way the problem is coded affects the efficiency and therefore $O(n^k)$. Problems that cannot be solved in a reasonable amount of time may be solved using an approximation algorithm or heuristics. But the solution found may be near optimal, but not the optimal solution. Some approximation algorithms can be applied to a wide range of problems and the rest are tailored to a specific problem.

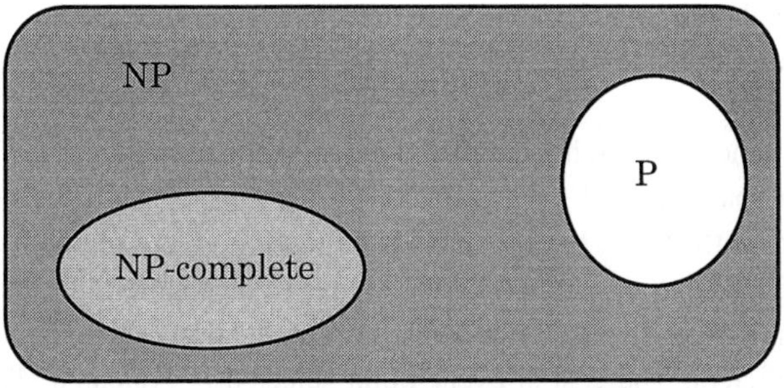

Figure 4.1: NP classification of complexity

Optimization problems (Van Laarhoven and Aarts 1987) can be formalised as a pair (R,C), where R is the finite - or countable infinite (on a computer) - set of configurations (the configuration space or the domain) and C is the cost function, $C{:}R \rightarrow R$, which assigns a real number to each configuration. R is the set of real numbers. The optimum value of C is C_{opt} and this occurs at the optimum configuration i_{opt}. If the objective is to minimize the cost then i_{opt} will satisfy:

$$C_{opt} = C\left(i_{opt}\right) = \min_{i \in R} C(i) \tag{4.1}$$

If the objective is to maximise the cost then i_{opt} will satisfy:

$$C_{opt} = C\left(i_{opt}\right) = \max_{i \in R} C(i) \tag{4.2}$$

There are three main types of search algorithms (Goldberg 1989): calculus-based, enumerative, and guided random. Calculus-based methods fall into two main categories: indirect and direct. Indirect methods seek local optima by solving a usually nonlinear set of equations resulting from setting the gradient of the cost function to zero. This method is tailored to a specific problem and will give the best solution. Direct methods (hill climbing) seek local optima by selecting a starting point and by moving a particular amount in a direction based on the local gradient (the steepest direction). This method will stop when it locates an optimal value (not necessarily the global optimal value) and, due to its discrete nature, it may not be able to locate the optimum value exactly, even if it is in the neighbourhood.

Enumerative schemes search the entire domain, evaluating the cost of every point in the search space. It will eventually locate the best point in the domain. This method is straightforward, but it is inefficient (and therefore impractical) for large domains. Guided random search schemes are based on enumerative techniques, but use additional information to guide the search. An entirely random search may be the most inefficient. The cost at every point selected is evaluated. This approach may never locate a reasonable solution and at worst it may evaluate more points than an enumerative search would have. More information about search techniques can be found in Filho and Treleaven (1994).

4.2 Genetic Algorithms

A GA is a heuristic search algorithm developed by Holland (1975). They fall into the guided random search category. The GA mainly differs from traditional search methods in the following areas (Chipperfield and Flemming 1996): (1) GAs search a population of points in parallel; (2) GAs use probabilistic transition rules, not deterministic ones; (3) GAs use an encoding of the points in the domain rather than points in the domain directly; and (4) GAs only need to be able to evaluate the cost function at each point in the domain (they do not need any other information about the problem, i.e. they are blind - they cannot see what is around them). The GA is robust and global (Goldberg 1989). The robustness implies that not only is the GA efficient in its operation, but it is also able to produce a solution, for a wide range of problems. It can provide a number of alternative solutions to a problem and can be useful in identifying alternative solutions to multiobjective problems.

The operation of GAs is similar to that of natural species. Organisms in nature live for a certain period, and during that period they produce offspring. In each successive generation the fittest survive and the weaker ones die off. The offspring produced by their parents inherit some, but not all, of the characteristics of their parents. They may be fitter or weaker, and their characteristics are not solely determined by their parents - other external effects may influence them. GAs are not the only idea borrowed from nature. Simulated annealing optimizes problems in a similar manner to the annealing process. Neural networks are fashioned along similar lines to brain cells. Fuzzy logic uses a more natural style of logic than the strict true/false reasoning.

GAs have been applied to solving various problems, such as the mapping problem (Talbi and Muntean 1993; Mansour and Fox 1992), graph partitioning (Muhlenbein 1991), communication network design (Coombs and Davis 1987), semiconductor layouts (Fourman 1985), control problems (Grefenstette 1989; Goldberg 1983; Michalewicz *et al.* 1990), function optimization (Bethke 1981), pattern recognition (Ankenbrandt *et al.* 1990). Robotics problems have been solved as well (Kwok and Sheng 1994; Chan and Zalzala 1993; Alander 1991; Solano and Jones 1994; Gill and Zomaya 1995b). Various other applications also exist.

GAs are used to optimize an objective function using a randomized, but guided, parallel iterative search of a finite discrete domain (Goldberg 1989; Holland 1975; Buckles *et al.* 1990), not a simulated reasoning process or random search. Within each iteration the search process simultaneously examines a set of strings. The number of strings is even as the crossover operation (explained later) operates on pairs of strings. Each string represents an individual point in the domain (search space). The coding of the string is arbitrary, but the mapping of the string into the domain must be unique.

The strings (**s**) consist of a series of n digits: $\mathbf{s} = \{b_1\ b_2\ ...,\ b_n\}$. The digits b are from an alphabet of size r. Therefore, the number of points in the domain to be searched is r^n. The string length should be selected such that the distance between two consecutive decoded strings in the domain is not significant, otherwise optimal values lying between two consecutive points will never be considered in the search. The most common representation of strings is the binary alphabet (which will be used in this work), i.e. $b_j \in \{0, 1\}$; $j = 1, 2, ..., n$. They have been shown, in general, to be optimal according to Goldberg (1989) and work well in solving problems in following chapters. For example, a variable x in the domain can be represented by the following binary string:

1 0 0 0 1 1 0 1 0 1 0 0 1 1 1 0 1 0 0 1

A vector can be represented by a subdividing a string into a set of concatenated substrings. The number of substrings is equal to the number of components of the vector. Each substring represents a component in the vector. The substrings do not have to be the same size, but in this work they are. A vector $\mathbf{x} = [x_1, x_2]$ can be represented by the following:

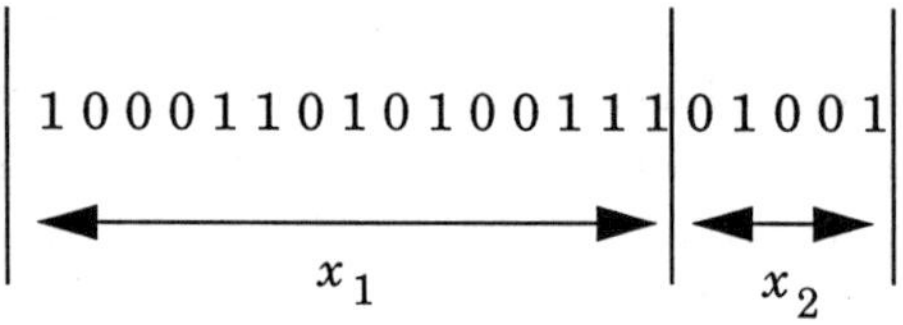

In coupled and nonlinear systems (such as robot manipulator kinematics) the portion of the individual contribution of each substring to the overall fitness (performance value) of the string is unknown. In this work, each string or substring is decoded to its numeric value and subsequently linearly scaled to suit the domain.

Let $\mathbf{z}_k$ denote the kth substring of string **s** (where there are j substrings, each of length s, therefore j.s = n), i.e. $\mathbf{s} = \{\mathbf{z}_1, \mathbf{z}_2, ..., \mathbf{z}_j\}$ and $\mathbf{z}_k = \{b_{(k-1)s+1}, b_{(k-1)s+2}, ..., b_{ks}\}$, where $k = 1, 2, ..., j$. Let **x** be a vector with each element containing the binary value of the corresponding substring from string **s**, i.e.

$$x_k = decode\left(z_k\right) = \sum_{l=1}^{\sigma} 2^{(\sigma - l - 1)} \times b_{((k-1) \times \sigma + l)} \tag{4.3}$$

The population of a GA (of size m) is the current set of strings, denoted $\mathbf{P}^t$, at iteration t ($t = 0, 1, ..., t_{\text{final}}$), i.e. $\mathbf{P}^t = \{\mathbf{s}^t_1, \mathbf{s}^t_2, ..., \mathbf{s}^t_m\}$. The initial population, $\mathbf{P}^0$, is selected randomly, without any bias, and subsequent generations, $\mathbf{P}^t$, are derived via a number of genetic operations applied to the previous population, $\mathbf{P}^{t-1}$. The main operations are: reproduction, crossover and mutation; and are sufficient for this case.

Each new generation should have a greater average fitness than the previous. If the population in the previous generation contains strings that have a greater fitness than the fittest strings in the current population then some of the old fit strings can be brought forward into the current population, replacing some of the new strings (either at random or replacing the worst). Figure (4.2) contains C-like pseudo code outlining the GA process.

The GA may never converge to the global maximum, but should obtain a reasonably fit solution in its locality. The fundamental theorem of GAs (schema theorem) demonstrates that the average fitness of the population should improve from generation to generation (Goldberg 1989). In some situations the GA may converge on a local maximum, instead of the global maximum, but the mutation operator will attempt to prevent this. The GA requires no auxiliary information about the fitness function (such as derivatives), but it must be an unconstrained maximization problem. The fitness function evaluates a particular strings performance. The fitness function does not have to be smooth, linear, differentiable, or noiseless, but should be able to be evaluated at each point in the domain, with a non-negative range. Problems involving minimization or having a negative cost can be converted to a suitable form for the GA via a mapping. The GA search should be terminated when a solution within the desired error margin is achieved or a set maximum number of iterations has been performed, and this should be less than:

$$\frac{r^n}{m} - 1 \tag{4.4}$$

```
counter = 0;
initialise_population(P(counter));
f(0) = evaluate_fitness(P(counter));
while (solution_not_found) {
          counter++;
          P(counter) = reproduce(P(counter-1));
          crossover(P(counter));
          mutate(P(counter));
          f(counter) = evaluate_fitness(P(counter));
          if (max(f(counter)) < max(f(counter-1)))
                 insert_previous_best(P(counter),P(counter-1));
}
```

Figure 4.2: GA pseudo-code

Otherwise the process is inefficient, since an enumerative search would cover the entire domain in the same time, without any repetition, and find the best possible solution.

4.2.1 Population Statistics

Fitness of string *i*:

$$f_i = fitness(\mathbf{x}_i) \tag{4.5}$$

Fitness sum:

$$\Psi = \sum_{i=1}^{m} f_i \tag{4.6}$$

Average population fitness:

$$f_{av} = \frac{\Psi}{m} = \frac{\sum_{i=1}^{m} f_i}{m} \tag{4.7}$$

Weight of string *i* (normalization):

$$w_i = \frac{f_i}{\Psi} \tag{4.8}$$

4.2.2 Reproduction Operator

The reproduction operator randomly selects new strings from the previous population proportional to their relative fitness. The fitter the string, the greater the chance of selection. Conversely, the weaker strings may not be selected at all. The process is based on roulette wheel selection (refer to Figure 4.3), each string is allocated a position on the wheel and the size of the segment is proportional to the fitness. *m* new strings are selected from the old population based on their relative fitness. A number, *y*, is randomly selected in (0,1]. If

$$\sum_{i=0}^{l-1} w_i < y \le \sum_{i=0}^{l} w_i, \text{ (with } w_0 \text{ set to 0)} \tag{4.9}$$

then string *l* is selected. Figure (4.4) contains C-like pseudo code outlining the operation.

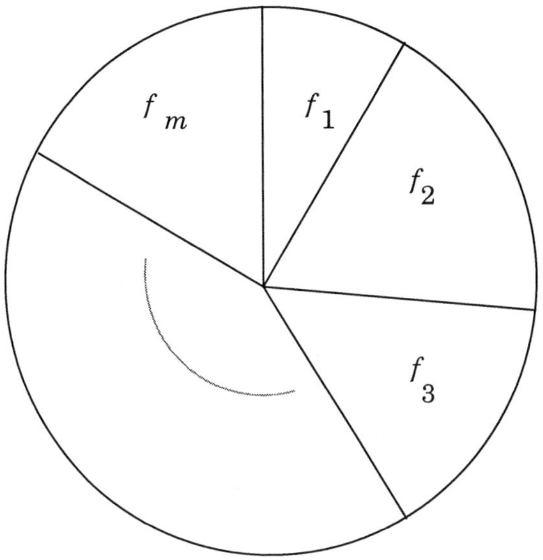

Figure 4.3: Roulette wheel selection

```
for (i = 0; i < population_size; i++) {
        element[i].value = decode(element[i].string);          /* (4.4) */
        element[i].fitness = fitness(element[i].value);        /* (4.5) */
        fitness_sum += element[i].fitness;                     /* (4.6) */
}
for (i = 0; i < population_size; i++) {
        element[i].weight = element[i].fitness / fitness_sum;/* (4.8) */
        if (i == 0) element[i].sum_weight = element[i].weight;
        else element[i].sum_weight = element[i].weight
              + element[i-1].sum_weight;
}
for (i = 0; i < population_size; i++) {
        y = random(0,1);
        counter = 0;
        for (j = 0; j < m; j++)
              if (element[j].sum_weight >= y) break;           /* (4.9) */
        new_element[i].string = element[j].string;
}
```

Figure 4.4: Reproduction pseudo-code

4.2.3 Crossover Operator

The crossover operation exchanges information between strings. It is only applied with a probability, p_c, to each pair of strings. Each string is paired off and the tail end of the two strings is swapped at a randomly chosen point (integer) from [1,*n*-1] if a randomly chosen number in (0, 1] for the pair is less than the crossover probability, p_c ($0 \le p_c \le 1$). Two binary strings below (Figure 4.5) have the crossover operation applied to them. The digits of the strings may also be shuffled to crossover at a random number of points. Figure (4.6) contains C-like pseudo code outlining the operation.

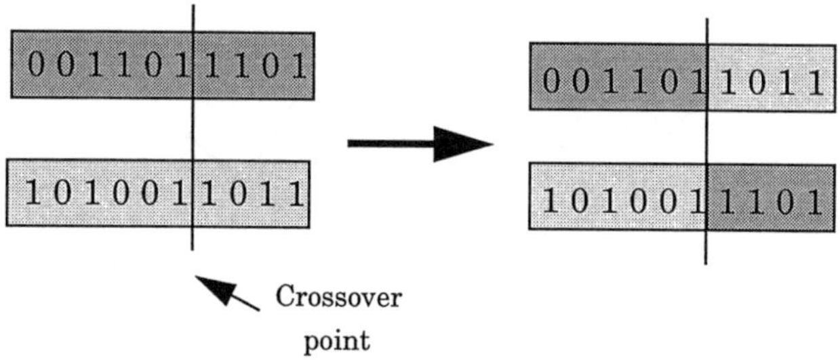

Figure 4.5: Crossover operation

```
for (i = 0; i < population_size; i += 2)
   if (random(0,1) < crossover_probability) {
      cross_over_point = random(1,string_length-1);
      for (j = cross_over_point; j < string_length; j++)
         exchange_bits(element[i].string[j],ele-
ment[i+1].string[j]);
```

Figure 4.6: Crossover pseudo-code

4.2.4 Mutation Operator

The mutation operator is used to randomly modify information in the strings. It is used to prevent the GA from converging on local maximum, rather than the global maximum, and can introduce new strings into the search. It is applied with a probability, p_m, to each digit. Each digit within each string is considered in turn, if a randomly chosen number in (0, 1] is less than the mutation probability, p_m ($0 \le p_m \le 1$), then that digit is inverted, i.e. $b'_{ij} = 1 - b_{ij}$. The mutation probability, p_m, should be very low, otherwise the GA will become nothing more than a random search. Figure (4.7) demonstrates the operation and Figure (4.8) contains C-like pseudo code outlining the operation.

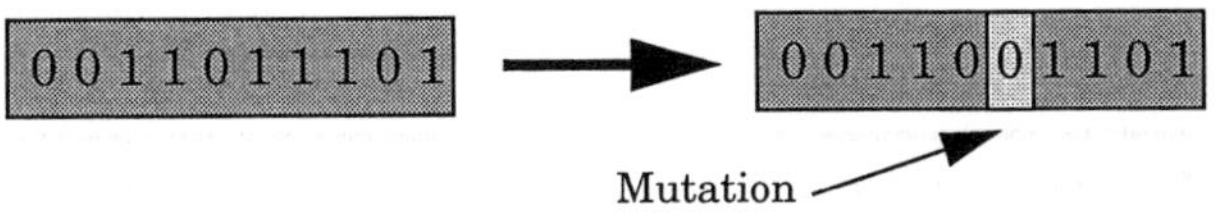

Figure 4.7: Mutation operation

```
for (i = 0; i < population_size; i++)
   for (j = 0; j < string_length; j++)
     if (random(0,1) < mutation_probability)
         element[i].string[j] = 1 - element[i].string[j];
```

Figure 4.8: Mutation pseudo-code

4.2.5 Genetic Algorithm Example

As an example, a GA has been used to optimize a bell shaped function to demonstrate its operation. The function is given in equation (4.10) and contains a global maximum at $(x, y) = (511.5, 511.5)$, the value of the function at this point is one. There are no minima and no other maxima. The variables x and y are directly mapped to 10-bit binary strings. The domain is the set of integers: $\langle x, y \mid x \in [0, 1023], y \in [0, 1023] \rangle$.

$$f(x, y) = \frac{1}{1 + \left(\frac{x}{102.3} - 5\right)^2 + \left(\frac{y}{102.3} - 5\right)^2} \tag{4.10}$$

Table (4.1) contains the GA parameters used. The selection of GA parameters is usually a trial and error process to find reasonable parameters which work for a particular problem. Figure (4.9) displays the progression of the average and maximum fitness at each iteration. The best value found is $f(x, y) = 0.971$ at $(x, y) = (527, 503)$, which has less than 3% error after 20 iterations. It only considered up to 420 points of the 1 048 576 point domain (i.e. approximately 0.04%). The most optimal value in this case could never be achieved since the point where it occurs lies between four points: (511, 511), (511, 512), (512, 512), and (512, 511). The best achievable value of the function with the coding used is 0.9995.

Population size	20
String length	20
Substrings	2
Iterations	20
Crossover probability	0.7
Mutation probability	0.001
Maximum number of strings retained	1

Table 4.1: GA example parameters

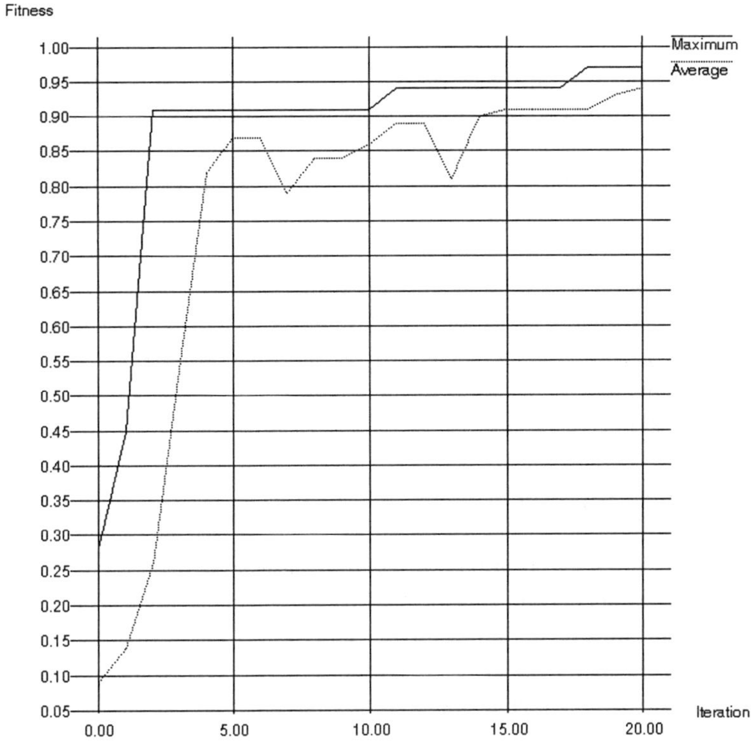

Figure 4.9: Fitness from GA example

4.2.6 Parallel Genetic Algorithms

GAs are parallel by nature. They search several points in the domain simultaneously. The GA can be easily implemented on a parallel architecture to speed up the processing of the operation (Grefenstette and Baker 1989; Tanese 1987). There three main ways to implement a GA in parallel (Goldberg 1989; Chipperfield and Fleming 1996):

- **Global GA**. It uses a master/slave approach. There is one master and it coordinates several slaves. The master performs all the operations except the fitness evaluations. The slaves handle the fitness evaluations. It is a coarse grained approach.

- **Migration GA** (also known as the network GA). In this case the population is divided between several independent GA processes. Each GA operates on its own subpopulation and occasionally shares some of its population (generally the fittest strings) with other GA processes. It is also considered to be a coarse grained approach.

- **Diffusion GA** (also known as the distributed, neighborhood, or cellular GA). In this case each string resides on its own processor, which performs all the genetic operations on it. Crossover occurs with strings in its locality and reproduction occurs within a pool of strings in the neighborhood. It is a fine grained approach.

4.2.7 Parallel Implementation

In this work the evaluation of the fitness function is significantly more computationally expensive than the GA itself. A global GA, using an asynchronous master/slave approach, has been used to compute the fitness of each string in parallel. The parallel GA was implemented on a transputer network. GAs have been implemented on various other architectures as well (Tanese 1987; Dorigo 1991; York *et al* 1994.). The approach allows interprocess communication by message passing as shown in Chapter (2).

The master stores the entire population, generates the initial population, and performs the reproduction operation. Each slave process receives a pair of strings from the master, decodes the strings, performs the crossover and mutation operations, and then evaluates the fitness of both strings. Each pair of new strings generated along with their associated fitness is returned back to the master to form the next population. The number of transputers in this case is less than half the population size. But, if the number of processors was greater than half the population size then it would be more efficient to have each slave process evaluate the fitness of a single string at a time.

A queueing system is used to distribute the pairs of strings to the slave processes, the operation is described in Chapter (2). Figure (4.10) contains C-like pseudo code outlining the operation.

```
in_counter = 0;                                      /* Number of pairs received */
out_counter = number_of_processors;                  /* Number of pairs sent */
population = reproduction(old_population);           /* Perform reproduction */
for (i = 0; i < number_of_processors; i++)
  send_data(i,element[i*2],element[i*2+1]);  /* Send data to processor */
while (in_counter < population_size / 2) {
  processor_number = recieve_data(element[in_counter*2],
     element[in_counter*2+1]);
           /* Receive data and ID code from ready processor */
  in_counter += 2;
  if (out_counter < population_size / 2) {
     send_data(processor_number,element[out_counter*2],
        element[out_counter*2+1]);
     out_counter += 2;
  }
}
```

Figure 4.10: PGA pseudo-code

4.3 Summary

This chapter has presented an introduction to GAs and how it has been implemented in this work. A brief outline of other search techniques has been presented. The GA has three main operators and only these have been used. Of the different ways of implementing a GA on a parallel architecture, only the global GA has been used as it is the most suitable for the problem. The work later in the thesis will use GAs to search for reasonable solutions for particular robotics problems. The example given demonstrates the operation of the GA in a simple case.

CHAPTER 5

INVERSE KINEMATICS

5.1 Introduction

There are two approaches used to analyse the motion of robot manipulators, they are the kinematics and kinetics (Craig 1986; Paul 1981; Schilling 1990; Snyder 1985; Yoshikawa 1990). The kinematics deal with the position, velocity and acceleration of a manipulator without considering the forces required to perform the motion. The kinetics, on the other hand, deal with the dynamics of the manipulator, i.e. the forces required to move it in a particular fashion. In path planning, this work is only concerned with the position of the manipulator at a particular point in time and not the forces required to realise the desired motion. The dynamics of robot manipulators is a separate problem and out of the scope of this work.

5.2 Denavit and Hartenberg Notation

A robot manipulator is represented by a series of links (Yoshikawa 1990; Craig 1986). Each link is a rigid object connected to the next link by a joint with a single degree of freedom. Each joint can be revolute (rotational) or prismatic (translational). A joint with two or more degrees of freedom can be modelled as a series of single degree of freedom joints interconnected by links of zero length. Link 0 is the base of the manipulator and the end-effector is attached to link n. Joint 0 and joint 1 are the same joints. A diagram of a robot manipulator with n links is shown in Figure (5.1).

The construction of a manipulator is shown in Figure (5.2). The **joint** is the connection between two links of the manipulator. It can either be revolute or prismatic. The motion of each joint is controlled by an actuator. The **link** is a rigid structure that is used to support the following links and any load carried. The **base** of the manipulator is often fixed, but may be mobile, it is the origin of the global reference frame. The **end-effector** (robot hand) is the final link of the manipulator and is attached to the manipulator at the wrist. It may be a tool that the manipulator uses in its operation. The **proximal links** are the links close to the base, while the

distal links are those that are the furtherest from the base. The **major axes** are the joints closest to the base and are used to position the manipulator. On the other hand, the **minor axes** are the joints furtherest from the manipulator (in the wrist) and are used to orient the end-effector.

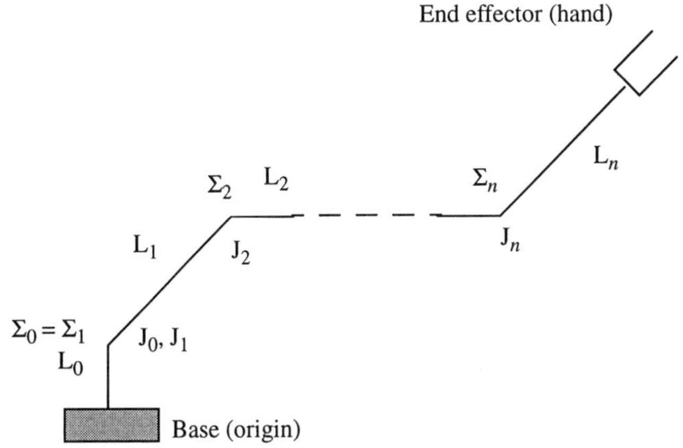

Notation: L_i = Link i, connecting joint i to joint i+1

J_i = Joint i, between link i-1 and link i.

Σ_i = Reference frame i, attached at joint i.

Figure 5.1: Manipulator model

The DOF of the manipulator is the number of independent joints or axes that it contains. In two-dimensions two DOF are required to position the end-effector and one extra DOF is required to orientate the end-effector. In three-dimensions three DOF are required to position the end-effector and three extra DOF are required to orientate the end-effector. If the DOF is greater than that required, then the redundancy can been used in obstacle and singularity avoidance or optimizing the end-effector trajectory (either time or energy minimization) (Yoshikawa 1990). A singularity occurs where a joint velocity must become infinite to maintain a constant cartesian velocity (Snyder 1985). The relationship between the joint velocity and cartesian velocity is described by the Jacobian matrix.

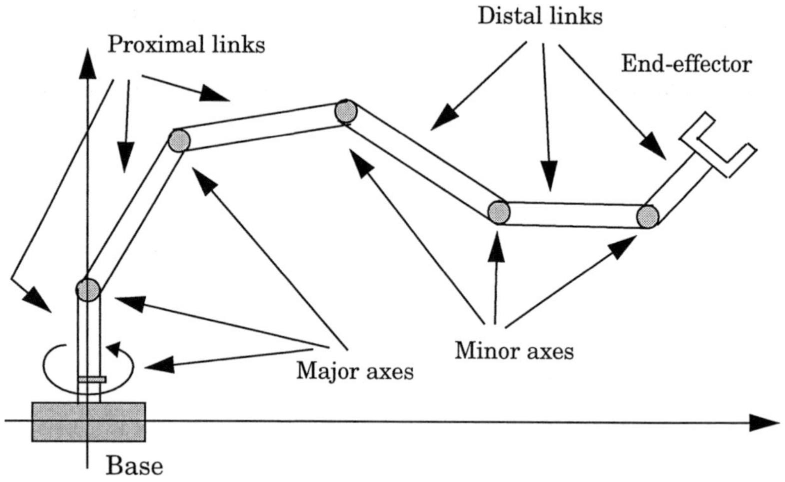

Figure 5.2: Robot manipulator

The mathematical model of link i is the common normal between joint axis i and joint axis i+1, as shown in Figure (5.3). The links are connected in ascending order from the base (link 0). Joint i connects link i-1 to link i. Each link has a right-handed coordinate frame associated with it. The ith coordinate frame (S_i) is attached to the ith link at the proximal end on joint i. The Z axis is the pivot and points towards the distal end of the manipulator. The X axis points towards the next axis and the Y axis is placed to make the coordinate frame a right-hand reference frame.

In the DH notation four parameters are used to describe each link (Denavit and Hartenburg 1955), they are (by definition) as follows. To describe the size and shape of link i (Yoshikawa 1990):

a_i = Link length (the distance along axis X_i from Z_i to Z_{i+1}).
α_i = Twist angle (the clockwise angle about axis X_i from Z_i to Z_{i+1}).

To describe the relative positional relation between links i-1 and i at joint i (Yoshikawa 1990):

d_i = Joint length (the distance along axis Z_i from X_{i-1} to X_i).
Θ_i = Joint angle (the clockwise angle about axis Z_i from X_{i-1} to X_i).

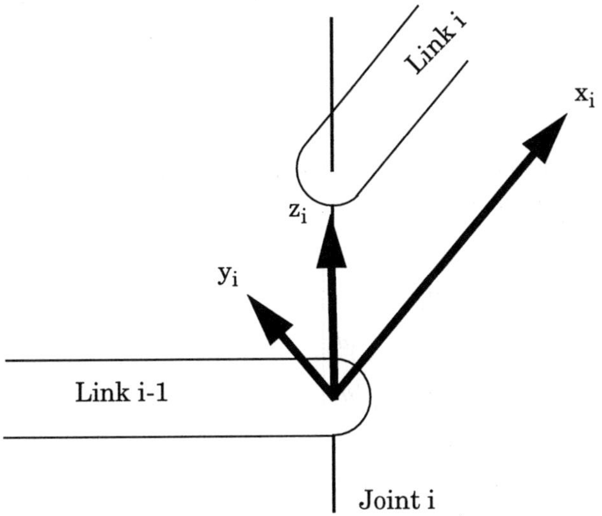

Figure 5.3: Link mathematical model

If link i is prismatic then the joint variable, q_i, is the joint length (d_i), otherwise, if the link is revolute then the joint variable is the joint angle (Θ_i).

For link i the origin of the coordinate frame Σ_i is located at the end-point of the mathematical model of link i on joint axis i. The homogenous transform describing the relation (which consists of four elementary rotations and translations) between Σ_i and Σ_{i-1} (i.e. it gives the position of reference frame i with respect to reference frame i-1) is (refer to (Yoshikawa 1990) for the derivation):

$$ {}^{i-1}\mathbf{T}_i = \begin{bmatrix} {}^{i-1}\mathbf{R}_i & {}^{i-1}\mathbf{p}_i \\ \mathbf{n} & s \end{bmatrix} \tag{5.1} $$

${}^{i-1}\mathbf{R}_i$ is a 3x3 rotation matrix of the ith moving coordinate frame attached to joint i:

$$ {}^{i-1}\mathbf{R}_i = \begin{bmatrix} \cos\Theta_i & -\sin\Theta_i & 0 \\ \cos(\alpha_{i-1})\sin\Theta_i & \cos(\alpha_{i-1})\cos\Theta_i & -\sin(\alpha_{i-1}) \\ \sin(\alpha_{i-1})\sin\Theta_i & \sin(\alpha_{i-1})\cos\Theta_i & \cos(\alpha_{i-1}) \end{bmatrix} \tag{5.2} $$

${}^{i-1}\mathbf{p}_i$ is a 3x1 column vector of the position of the origin of the ith coordinate frame relative to coordinate frame i-1.

$$ {}^{i-1}\mathbf{p}_i = \begin{bmatrix} a_{i-1} \\ -\sin(\alpha_{i-1})d_i \\ \cos(\alpha_{i-1})d_i \end{bmatrix} \tag{5.3} $$

$\mathbf{n}$ is a 1x3 perspective vector, usually zero; and s is a scaling factor, usually one.

A 4-vector defined with respect to reference frame A is:

$$ {}^{A}\mathbf{v} = [{}^{A}x, {}^{A}y, {}^{A}z, 1]^{T} $$

The relationship between the two vectors, ${}^{A}\mathbf{v}$ and ${}^{B}\mathbf{v}$, in reference frames A and B, respectively, is:

$$ {}^{A}\mathbf{v} = {}^{A}\mathbf{T}_{B}\,{}^{B}\mathbf{v} \tag{5.4} $$

Note that: $$ {}^{A}\mathbf{T}_{C} = {}^{A}\mathbf{T}_{B}\,{}^{B}\mathbf{T}_{C} \tag{5.5} $$

On a manipulator the base (link 0) is the global reference frame. To relate any link reference frame S_k of link k to the base S_0, the following transformation is used:

$$ {}^{0}\mathbf{T}_{k} = {}^{0}\mathbf{T}_{1}\,{}^{1}\mathbf{T}_{2} \ldots {}^{k-1}\mathbf{T}_{k} \tag{5.6} $$

The coordinates of the origin in reference frame k (i.e. the position of joint k) with respect to the base are:

$$ {}^{0}\mathbf{j}_{k} = {}^{0}\mathbf{T}_{k}\,{}^{k}\mathbf{j}_{k} \tag{5.7} $$

Where: ${}^{0}\mathbf{j}_{k} = [{}^{0}x_{k}, {}^{0}y_{k}, {}^{0}z_{k}, 1]$, and

${}^{k}\mathbf{j}_{k} = [{}^{k}x_{k}, {}^{k}y_{k}, {}^{k}z_{k}, 1] = [0, 0, 0, 1]$.

The forward kinematics equation determines the position of the end-effector ($\mathbf{r} = [r_1, r_2, ..., r_m]^T$) uniquely from the joint variables ($\mathbf{q} = [q_1, q_2, ..., q_n]^T$), where $m \leq n$ and n is the DOF (or number on links/joints). The forward kinematics is a transformation from configuration space to cartesian space. It is solved using a matrix multiplication of the form in (5.6) to relate the coordinates of the end-effector to the global reference frame. The coordinates of all the joints can be found in this manner, and are given by:

$$ \mathbf{r} = \mathbf{f}(\mathbf{q}) \tag{5.8} $$

Conversely, the inverse kinematics equation determines the joint variables from the end-effector position, a transformation from cartesian space to configuration space, and is given by:

$$ \mathbf{q} = \mathbf{f}^{-1}(\mathbf{r}) \tag{5.9} $$

Unlike the forward kinematics problem (which can be solved using (5.6)), the inverse kinematics problem may be difficult to solve. It involves the inversion of non-linear transcendental equations. This is usually done using iterative numerical techniques or closed-form solutions using algebraic or geometric methods. Closed-form solutions do not consider constraints placed on the manipulator (such as joint limits). The solution to the inverse kinematics is dependent on the structure of the manipulator and is therefore applicable to a single manipulator. In some cases there may not be a solution or more than one solution may exist. In cases where there is redundancy $m < n$, there may be an infinite number of solutions (degeneracy). There is no redundancy if $m = n$. If neither avoidance or optimization is required then geometric methods may be used. In this case a heuristic search has been used (Section 5.3).

5.3 GA Search

The GA is used to search for a suitable joint vector to realize the desired end-effector position (within a particular error margin). The use of the GA in this manner is similar to Parker *et al.* (1989). But, in this case, the GA searches in a restricted portion of the joint extremities, instead of between the joint extremities themselves, to simplify the calculation of the fitness function. The path followed by the end-effector is discrete, in each step a maximum distance of δd can be moved by the end-effector during time interval δt. During the time interval δt the joints of the manipulator can move a maximum of δq. The search for the required joint angles takes place from the current joint positions $\pm\delta q$.

Initially, the target is the starting point and the search takes place between joint extremities, or initial starting joint variables for the manipulator can be specified. If the initial joint angles are not specified then the population of the GA may have to be larger and it will need a longer time to find a suitable set of starting angles, as the search takes place over a much larger domain. The string lengths may also need to be increased to compensate for the loss in accuracy due to searching a larger domain.

After the initial configuration is selected or located the position of the end-effector is computed. From this initial position the next end-effector position is selected (at a distance δd from the previous position). The GA is then used to search for suitable joint angles for the manipulator to realize this new position. The process then continues until the goal has been reached. Figure (5.4) contains a block diagram outlining the operation.

In this work configuration space is not used and a path will refer to the end-effector path through cartesian space. The trajectory is the corresponding joint positions required to realize the path.

Each genetic string is divided into a set of (equal length) substrings, each substring corresponds to a particular joint variable. The substrings are linearly transformed into joint variables which are then applied to the forward kinematics equation. If a particular joint is revolute with a full $360°$ rotation then the resulting angle found is adjusted (by adding or subtracting $360°$) to ensure it lies between $0°$ and $360°$. The fitness is evaluated in terms of how close the end-effector is to the desired position.

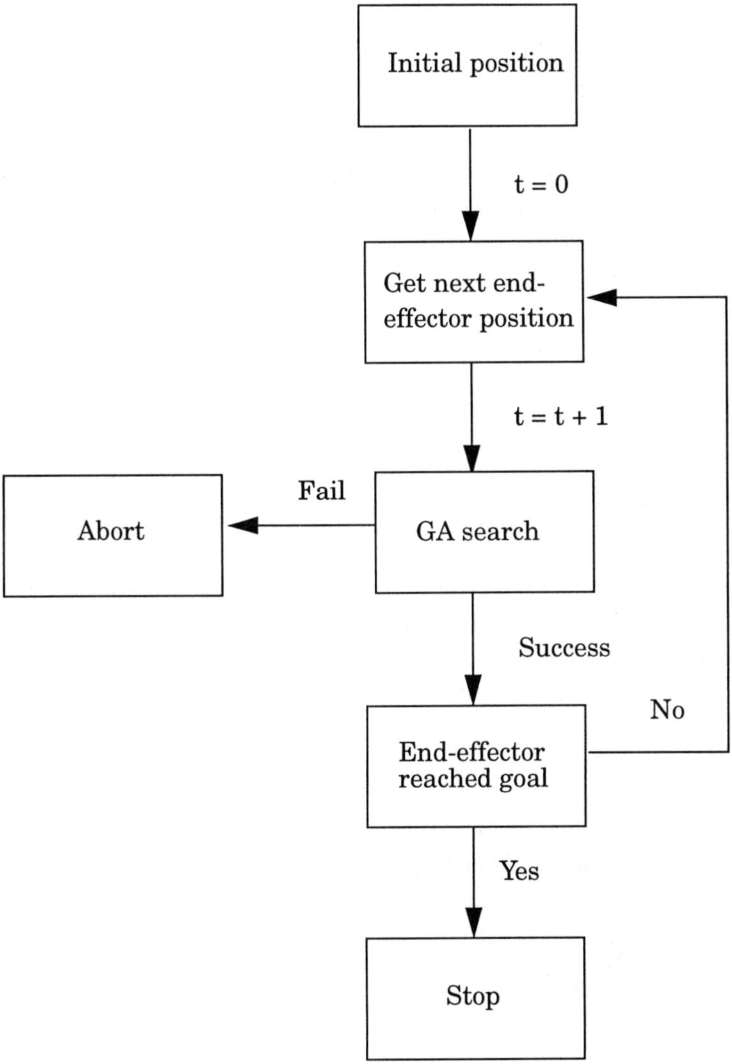

Figure 5.4: GA Kinematics procedure

The linear mapping from each genetic substring to a joint variable is as follows, for joint *k*:

$$q'_k = \left(\frac{q_{kx} - q_{kn}}{2^{\sigma} - 1}\right) \times x_k + q_{kn} \tag{5.10}$$

The joint limits (q_{kx} and q_{kn}) at the start are the maximum and minimum allowable joint positions, during the motion they are defined as:

Maximum limit: $q_{kx} = q_k + \delta q_k$ (5.11)

Minimum limit: $q_{kn} = q_k - \delta q_k$ (5.12)

The fitness function is defined as:

$$f = \frac{1}{1 + w_1 \times e_1} \tag{5.13}$$

The fitness function (5.13) is defined this way so that as the error increases from zero to infinity, the fitness drops from one to zero. This is needed because the GA attempts to maximize an objective function, not minimize it. It also places a finite limit on the fitness for any possible error.

The positioning error is: $e_1 = \frac{\|\mathbf{r}' - \mathbf{r}\|}{2 \times R}$ (5.14)

The maximum reach of the manipulator is the total length of the mathematical model of the manipulator and in defined as:

$$R = \sum_{i=1}^{\varphi} \sqrt{d_i^2 + a_i^2} \tag{5.15}$$

Where: q_k = Current joint variable *k*,
q'_k = New joint variable *k*,
δq_k = Maximum allowable joint motion for joint *k*,
x_k = Decoded value of genetic substring *k* (4.3),
s = Substring length,
j = Number of substrings or links (DOF),

w_1 = Error weight,
$\mathbf{r}$ = Actual end-effector position,
$\mathbf{r}'$ = Desired end-effector position,
d_i = Joint length (Denavit and Hartenburg 1955), and
a_i = Link length (Denavit and Hartenburg 1955).

5.4 Results and Examples

Two cases are examined to demonstrate the GA path planner. For the first example, a two DOF planar manipulator is used. Both joints are revolute, the first link is two units long and the second is one unit long. The base is at (0,0) and the straight-line path contains 100 samples, with the end-effector starting from (2,1) and finishing at (1,2). Figure (5.5) shows the manipulator at each tenth sample. A summary of the GA parameters is in Table (5.1), and Table (5.2) contains a summary of the results, the error is calculated using the positioning error from (5.14). Figures (5.6), (5.7) and (5.8) are graphs of the number of iterations, error, and fitness, respectively, at each sample.

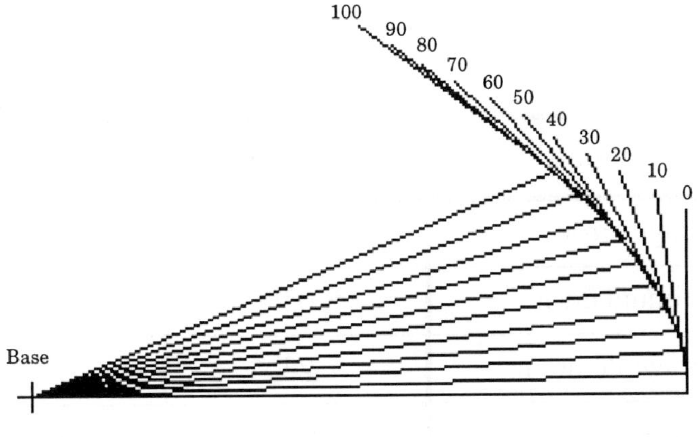

Figure 5.5: Planar manipulator with GA

Parameters	Example 1	Example 2
Population size	20	20
String length	40	40
Maximum iterations	200	200
Crossover probability	0.7	0.7
Mutation probability	0.01	0.01
Target error	0.2%	0.5%

Table 5.1: GA parameters

Results	Example 1	Example 2
Average error	0.15%	0.41%
Maximum error	3.03%	0.755%
Average iterations	23	28
Success rate	97%	97%

Table 5.2: Summary of results

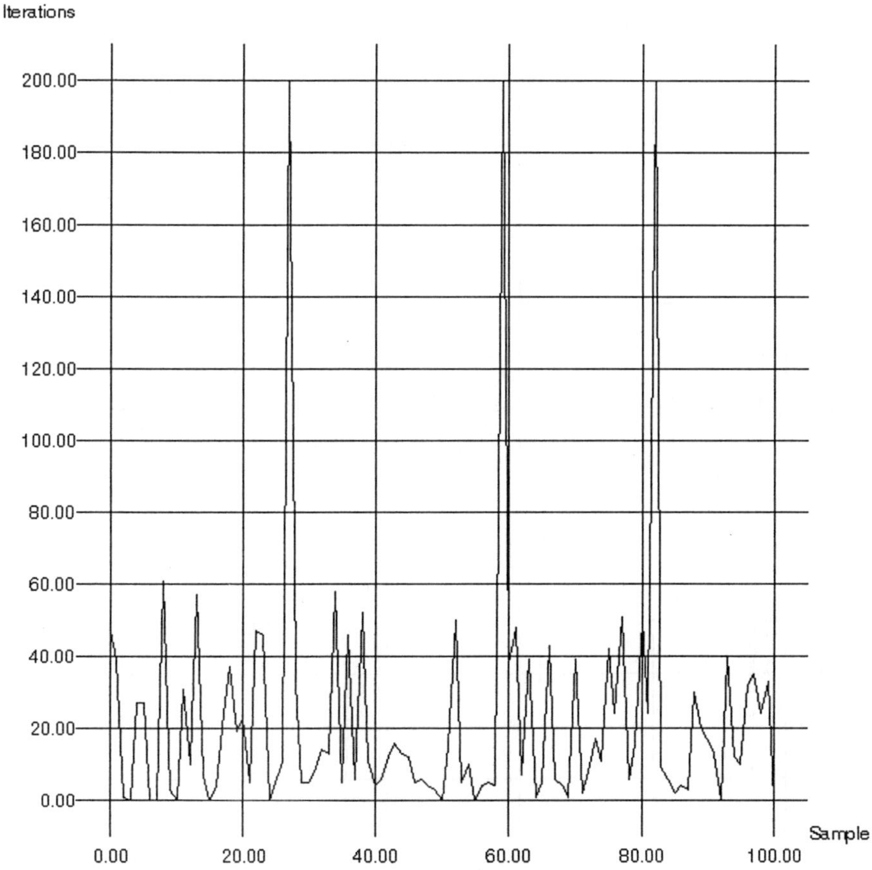

Figure 5.6: GA iterations

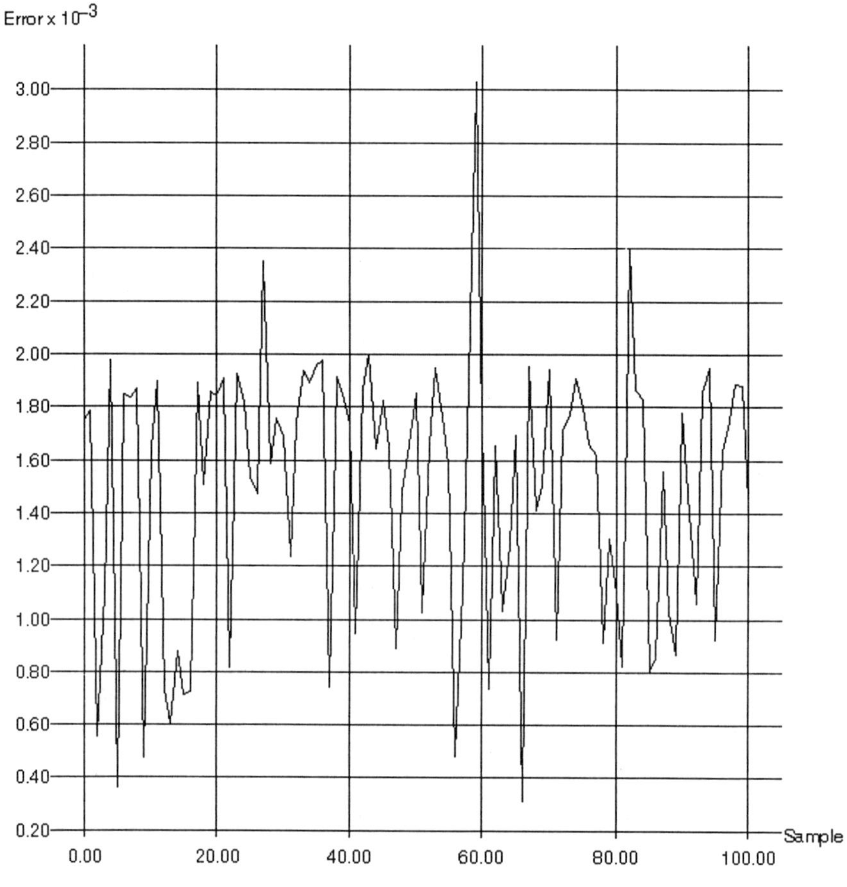

Figure 5.7: GA error

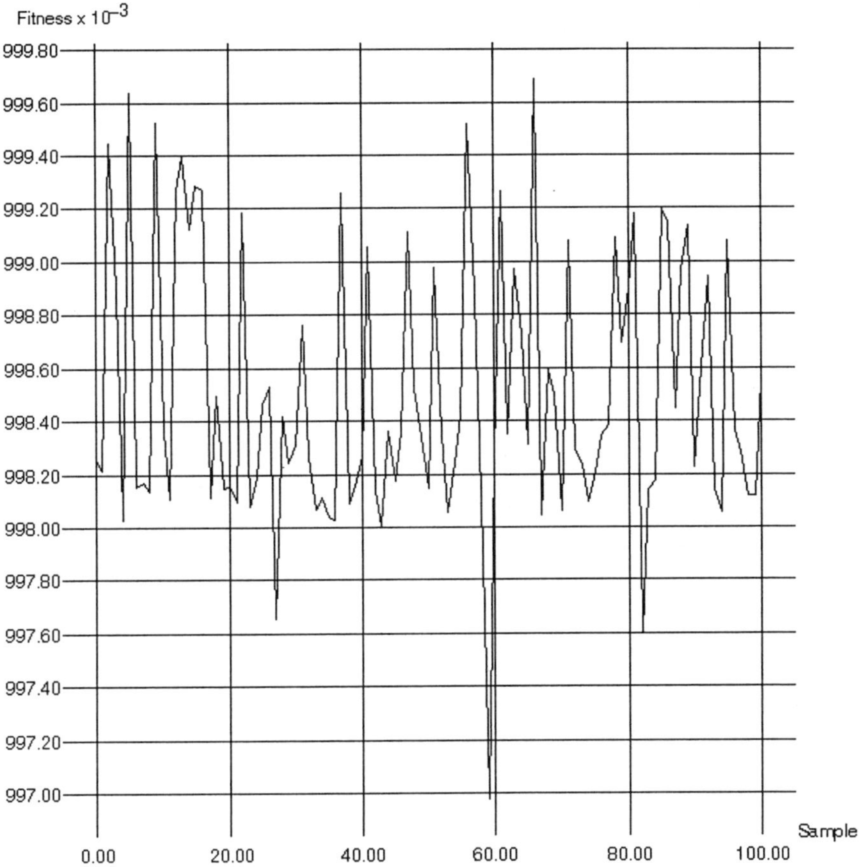

Figure 5.8: GA fitness

For the second example, a three DOF manipulator is used, in a three-dimensional environment. All the joints are revolute, and the first link is zero units long, the second is two units long, and the third is one unit long. The base is at (0,0) and the straight-line path contains 100 samples, with the end-effector starting from (2, 2, 0) and finishing at (-1.5, 1.5, 1). Figure (5.9) shows the manipulator at each tenth sample. A summary of the GA parameters is in Table (5.1), and Table (5.2) contains a summary of the results. Figures (5.10), (5.11) and (5.12) are graphs of the number of iterations, error, and fitness, respectively, at each sample.

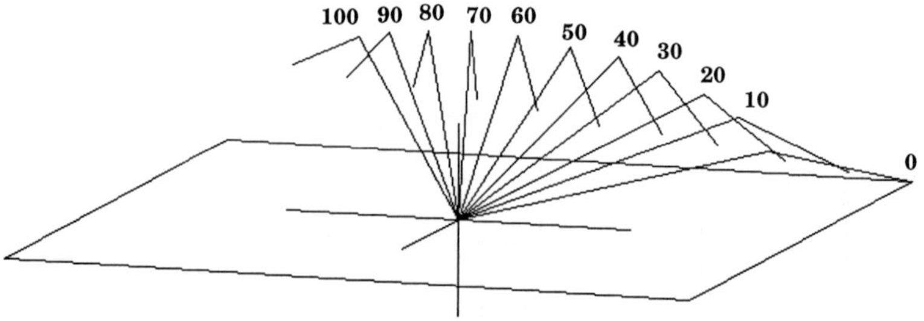

Figure 5.9: Manipulator with GA

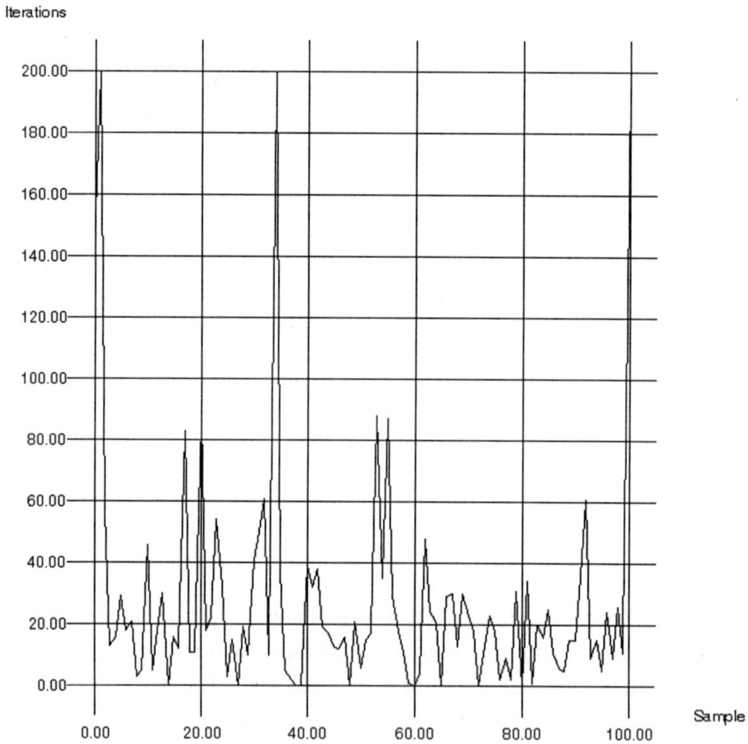

Figure 5.10: GA iterations

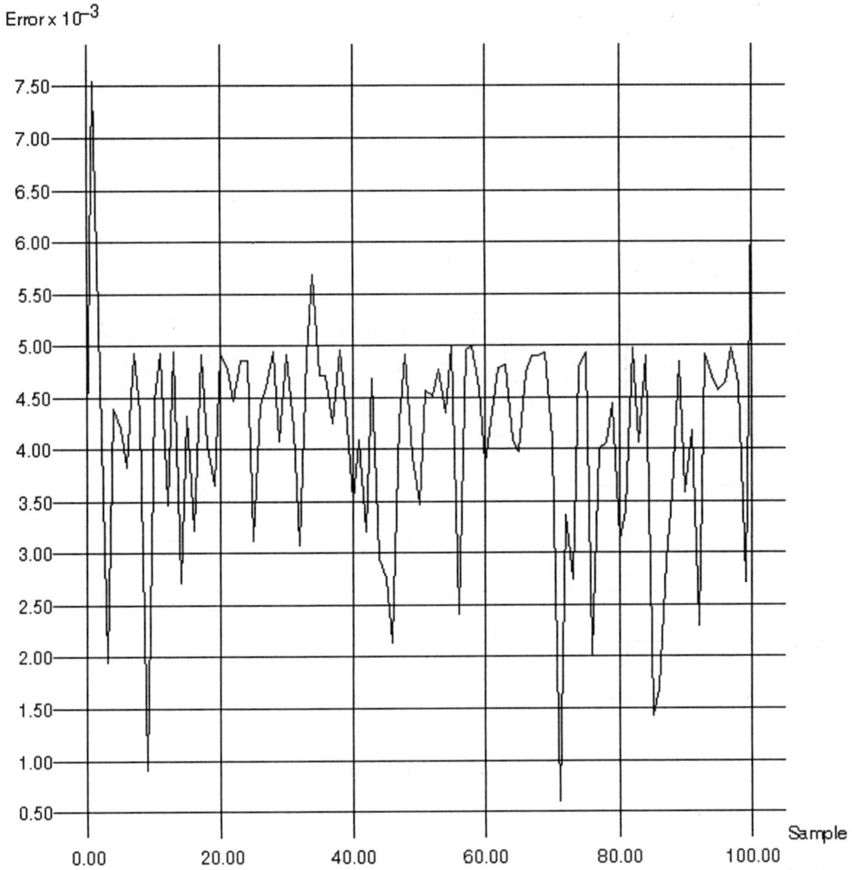

Figure 5.11: GA error

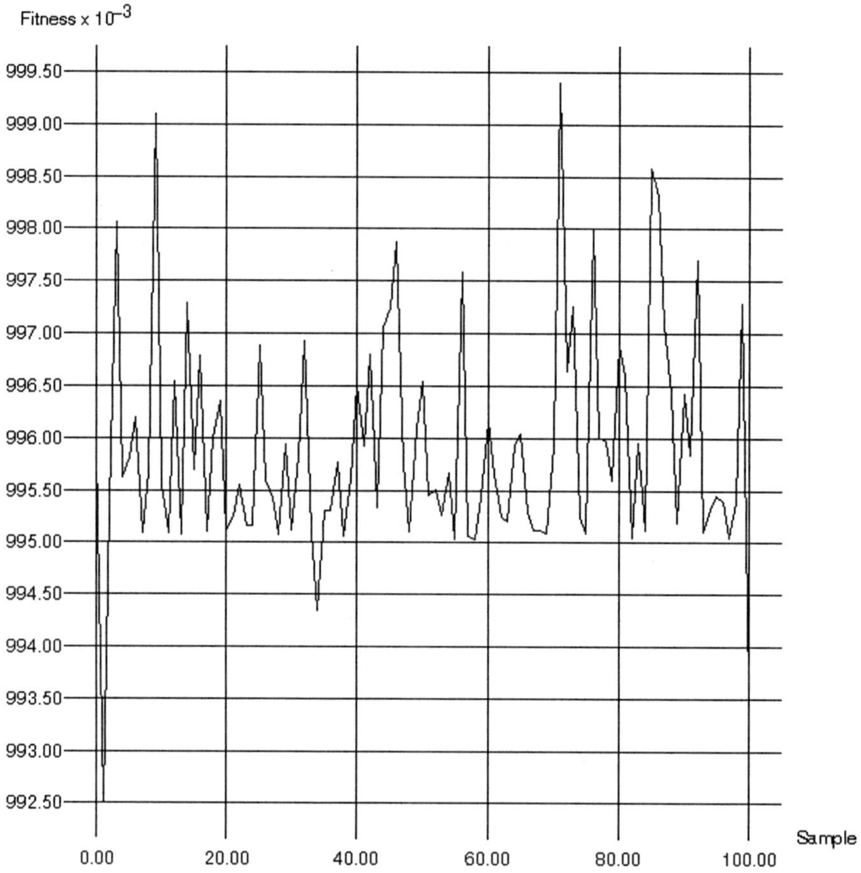

Figure 5.12: GA fitness

5.5 Conclusions

The inverse kinematics problem is one of numerous difficult robotics problems. Initially the structure and mathematical model of a manipulator was examined. Then a GA was implemented to solve the inverse kinematics problem. The results demonstrated that the inverse kinematics problem can be solved, using a GA, within particular error limits. In Chapter 7 the error term in the fitness function will be extended to allow for collision avoidance.

CHAPTER 6

COLLISION DETECTION

6.1 Introduction

Collision detection is an important part of path planning for robot manipulators. A computer simulation can be used to test manipulator trajectories for possible collisions, before being used in a real situation. Collisions can occur between a manipulator and other manipulators or obstacles in the workspace.

Various methods have been proposed in the past to detect collisions involving manipulators. Some previous work has considered only detecting collisions between two manipulator wrists in the workspace (Basta *et al.* 1988; Mehrotra *et al.* 1989), which does not consider the rest of the manipulator or any obstacles in the workspace. Other methods are more complex by detecting intersections between arbitrary objects (Culley 1986; Cameron 1990; Schweikard 1991; Gilbert *et al.* 1988; Myers 1985; Bonney *et al.* 1983; Cameron 1989), which have a very high degree of accuracy. However, the highly accurate methods are more computationally intensive, and hence slower when such accuracy is not required, and a simplified model would be sufficient.

The method developed in this work (Gill and Zomaya 1995a) treats the problem as follows. Manipulators are treated as a series of line segments (links) with an associated thickness and objects are modelled as a series of connected plane segments (surfaces) or spheres. The collision detection routine involves computing the minimum distance between two line segments, a line segment and a plane, or a line segment and a sphere. Two methods can be used to derive the minimum distances. One method uses geometry and linear algebra (Daniel 1981), and the other uses calculus. Both methods give exactly the same results, but only the first method will be used here. The calculus approach minimizes the distance between two arbitrary points with respect to the parametric variable(s). If the computed minimum distance is less than a particular threshold, based on the link thicknesses, then a collision is said to have occurred.

The methods used are discrete, i.e. they detect collisions only at discrete intervals of time. So the sampling rate of the trajectory must be set appropriately so collisions are not missed.

6.2 Modelling of Manipulator and Obstacles

The model of each link is based on the mathematical model of the link which is treated as a capped cylinder, with hemispherical caps centered on each of the two connected joints. The cylinder and caps have a radius *r*, as shown in Figure (6.1). The radius, *r*, is chosen so that the link approximates as closely as possible the real link, as real links do not necessarily have this shape. The links are treated as line segments. To detect if two links intersect each other compute the minimum distance between them, if this distance is less than the sum of the radii of both links then they intersect. To detect if a collision has occurred between two manipulators, test each link on one manipulator against all links on the other manipulator for any intersections. When computing the distance between two line segments, different formulas are used for parallel and non-parallel line segments, so whether or not the two line segments are parallel must be determined first.

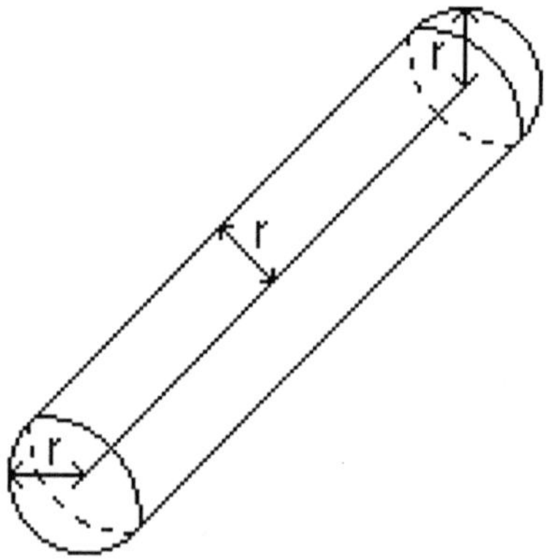

Figure 6.1: A robot link

Link i of a manipulator is represented by a parametric equation (refer to Figure 6.2) with the end-points being joints i (${}^{0}\mathbf{j}_i$) and i+1 (${}^{0}\mathbf{j}_{i+1}$). The parametric equation of a line segment connecting these two points is:

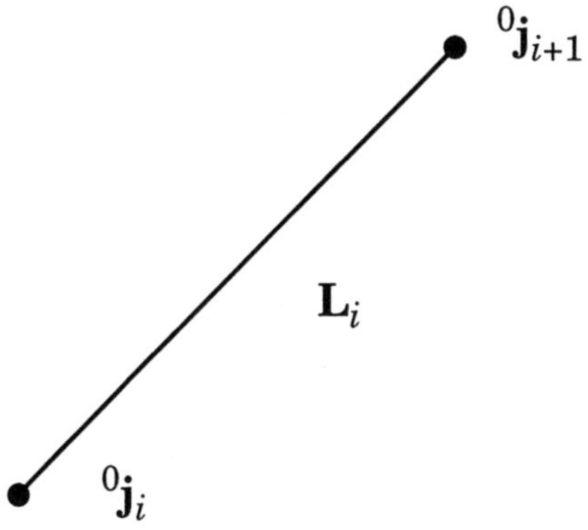

Figure 6.2: The parametric representation of link i

$$\mathbf{L'} = \mathbf{L'}(t) = [L'_x, L'_y, L'_z] = \mathbf{L} + t\,\mathbf{v} = [L_x, L_y, L_z] + t\,[v_x, v_y, v_z] \tag{6.1}$$

Starting point: $\mathbf{L} = [L_x, L_y, L_z] = {}^{0}\mathbf{j}_i$,

Direction vector: $\mathbf{v} = [v_x, v_y, v_z] = {}^{0}\mathbf{j}_{i+1} - {}^{0}\mathbf{j}_i$, and

Parametric variable: $0 \le t \le 1$.

Each object is described by a series of plane segments or spheres. To detect if a collision has occurred between a manipulator and an object: Compute the minimum distances between each link of the manipulator and each plane segment of the object or the sphere. If any of the distances are less than the radius of the link involved then a collision has occurred.

A plane is described by the equation:

$$a\,(x - x_0) + b\,(y - y_0) + c\,(z - z_0) = a\,x + b\,y + c\,z + d = 0 \tag{6.2}$$

Where: $d = -(a\, x_0 + b\, y_0 + c\, z_0)$ (6.3)

Note that the vector $[a, b, c]$ is normal to the plane, as in Figure (6.3).

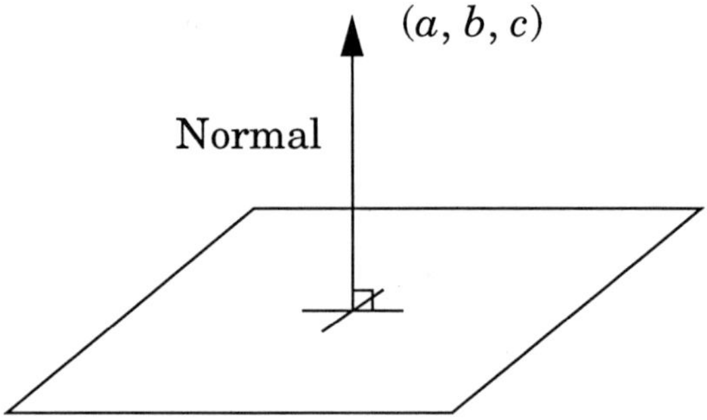

Figure 6.3: A plane segment

A plane segment has m corner points ($m > 2$):

$$[x_1, y_1, z_1], [x_2, y_2, z_2], \ldots, [x_m, y_m, z_m].$$

The equation of this plane is obtained by setting:

$$[x_0, y_0, z_0] = [x_1, y_1, z_1], \text{ and}$$

$$[a, b, c] = [x_1 - x_2, y_1 - y_2, z_1 - z_2] \times [x_1 - x_3, y_1 - y_3, z_1 - z_3] \quad (6.4)$$

i.e.
$$a = (y_2 - y_1)(z_3 - z_1) - (z_2 - z_1)(y_3 - y_1),$$
$$b = (z_2 - z_1)(x_3 - x_1) - (x_2 - x_1)(z_3 - z_1),$$
$$c = (x_2 - x_1)(y_3 - y_1) - (y_2 - y_1)(x_3 - x_1), \text{ and}$$
$$d = -(a\, x_1 + b\, y_1 + c\, z_1).$$

6.3 Preliminary Calculations

This section will present the main calculations used in the collision detection routine. Initially, the task of computing the distances between lines and spheres will be considered. Then computing the distances between parallel and non-parallel line segments will be considered. Finally, the distances between line segments and plane segments will be computed.

6.3.1 Preliminary Calculations for Lines

An arbitrary point in space is denoted: $\mathbf{p} = [p_x, p_y, p_z]$

The distance between two points, $\mathbf{p}_1 = [p_{x1}, p_{y1}, p_{z1}]$ and $\mathbf{p}_2 = [p_{x2}, p_{y2}, p_{z2}]$, is:

$$d = \sqrt{\left(p_{x1} - p_{x2}\right)^2 + \left(p_{y1} - p_{y2}\right)^2 + \left(p_{z1} - p_{z2}\right)^2} \tag{6.5}$$

For two lines: $\mathbf{L'}_1 = \mathbf{L}_1 + t\,\mathbf{v}_1$ and $\mathbf{L'}_2 = \mathbf{L}_2 + s\,\mathbf{v}_2$, define:

$$\alpha = \mathbf{v}_1 \cdot \mathbf{v}_1 = v_{x1}^2 + v_{y1}^2 + v_{z1}^2 \tag{6.6}$$

$$\beta = \mathbf{v}_1 \cdot \mathbf{v}_2 = v_{x1}v_{x2} + v_{y1}v_{y2} + v_{z1}v_{z2} \tag{6.7}$$

$$\gamma = \mathbf{v}_2 \cdot \mathbf{v}_2 = v_{x2}^2 + v_{y2}^2 + v_{z2}^2 \tag{6.8}$$

$$\Delta = \begin{vmatrix} \alpha & -\beta \\ -\beta & \gamma \end{vmatrix} = \alpha\gamma - \beta^2 \tag{6.9}$$

$$\delta = v_{x1}\left(L_{x1} - L_{x2}\right) + v_{y1}\left(L_{y1} - L_{y2}\right) + v_{z1}\left(L_{z1} - L_{z2}\right) \tag{6.10}$$

$$\omega = v_{x2}\left(L_{x1} - L_{x2}\right) + v_{y2}\left(L_{y1} - L_{y2}\right) + v_{z2}\left(L_{z1} - L_{z2}\right) \tag{6.11}$$

If D = 0 then lines $\mathbf{L'}_1$ and $\mathbf{L'}_2$ are parallel, for proof refer to Appendix (1).

6.3.1.1 Point and a Line

The distance (as a function of t) between a point, $\mathbf{p} = [p_x, p_y, p_z]$, and an arbitrary point, $[L_x + t\,v_x, L_y + t\,v_y, L_z + t\,v_z]$, on a line, $\mathbf{L}' = \mathbf{L} + t\mathbf{v}$, is:

$$d(t) = \sqrt{\left(L_x + tv_x - p_x\right)^2 + \left(L_y + tv_y - p_y\right)^2 + \left(L_z + tv_z - p_z\right)^2} \quad (6.12)$$

To compute the minimum distance between the point and the line (refer to Figure 6.4), a point, t_m, must be found on the line such that $d(t_m)$ is the minimum distance. The line spanning the minimum distance between the point and the line is normal to the line, therefore:

$$\mathbf{n} \cdot \mathbf{v} = n_x\, v_x + n_y\, v_y + n_z\, v_z = 0 \quad (6.13)$$

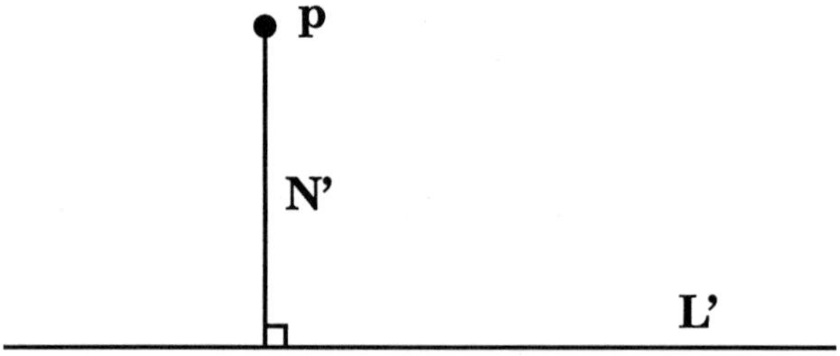

Figure 6.4: Minimum distance between a point and a line segment

Where the equation of the normal line is:

$$\mathbf{N}' = \mathbf{N}'(r) = [N'_x, N'_y, N'_z] = \mathbf{N} + r\,\mathbf{n} = [N_x, N_y, N_z] + r\,[n_x, n_y, n_z] \quad (6.14)$$

If $\mathbf{N}$ is set to equal $\mathbf{p}$ (i.e. the point $\mathbf{p}$ is the starting point of the normal), then: $N_x = p_x$, $N_y = p_y$, and $N_z = p_z$.

The intersection between the normal line and the line occur at:

$$[p_x + r\,n_x, p_y + r\,n_y, p_z + r\,n_z] = [L_x + t\,v_x, L_y + t\,v_y, L_z + t\,v_z] \quad (6.15)$$

Therefore, from the components of Equation (6.15):

$$r\,n_x - t\,v_x = L_x - p_x \tag{6.16}$$
$$r\,n_y - t\,v_y = L_y - p_y \tag{6.17}$$
$$r\,n_z - t\,v_z = L_z - p_z \tag{6.18}$$

From Equations (6.13), (6.16) - (6.18):

$$\begin{bmatrix} r & 0 & 0 & -v_x \\ 0 & r & 0 & -v_y \\ 0 & 0 & r & -v_z \\ v_x & v_y & v_z & 0 \end{bmatrix} \begin{bmatrix} n_x \\ n_y \\ n_z \\ t \end{bmatrix} = \begin{bmatrix} L_x - p_x \\ L_y - p_y \\ L_z - p_z \\ 0 \end{bmatrix} \tag{6.19}$$

$$\therefore t_m = \frac{v_x\left(p_x - L_x\right) + v_y\left(p_y - L_y\right) + v_z\left(p_z - L_z\right)}{v_x^2 + v_y^2 + v_z^2} \tag{6.20}$$

If **L'** = **L'**$_1$ and **p** = **L'**$_2$ (with $\mathbf{v}_2 = \mathbf{0}$) then t_m = -d/a, or
if **L'** = **L'**$_2$ and **p** = **L'**$_1$ (with $\mathbf{v}_1 = \mathbf{0}$) then t_m = w/g.

6.3.1.2 Point and a Line Segment

Compute t_m as in the Section (6.3.1.1) and if the parametric variable, *t*, of line **L'** is restricted to the interval [0, 1] (i.e. line **L'** is now a line segment), then t_m must also lie in the interval [0, 1], i.e.

If $t_m < 0$ then $t_m = 0$, or
If $t_m > 1$ then $t_m = 1$.

6.3.1.3 Sphere and Line Segment

To compute the distance between a sphere and a line segment, compute the distance between the centre point and the line segment using Section (6.3.1.2). Subtract the radius of the sphere from the result.

6.3.1.4 Two Non-Parallel Lines

The distance (a function of s and t) between two arbitrary points on each line ($\mathbf{L'}_1$ and $\mathbf{L'}_2$) is:

$$d = d(s, t) = |\boldsymbol{n}| = \sqrt{n_x^2 + n_y^2 + n_z^2} \tag{6.21}$$

Where: $n_x = L_{x1} + t\, v_{x1} - L_{x2} - s\, v_{x2}$,
$n_y = L_{y1} + t\, v_{y1} - L_{y2} - s\, v_{y2}$, and
$n_z = L_{z1} + t\, v_{z1} - L_{z2} - s\, v_{z2}$.

The line ($\mathbf{N'} = \mathbf{N} + r\mathbf{n}$) spanning the minimum distance between the two lines is normal to both of the lines (refer to Figure 6.5). The end points of the normal spanning the distance are set to a point on each of the two lines, i.e.

At $r = 1$, the normal intersects $\mathbf{L'}_1(t)$ at $\mathbf{L'}_1(t_m)$, and
At $r = 0$, the normal intersects $\mathbf{L'}_2(s)$ at $\mathbf{L'}_2(s_m)$.

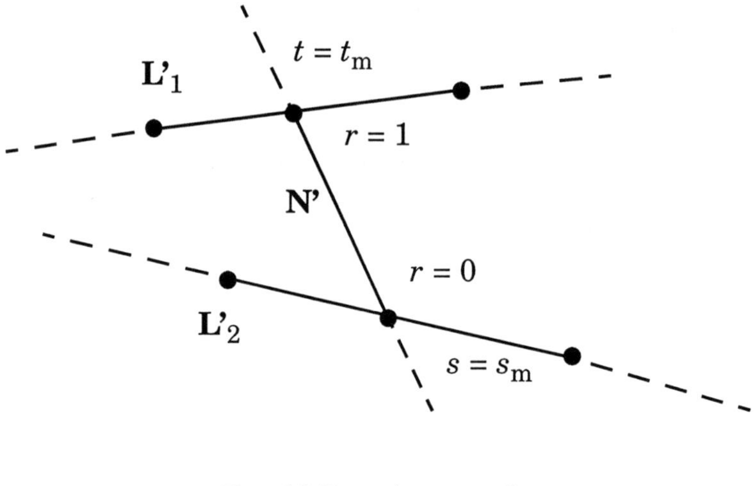

Figure 6.5: Distance between two lines

At $r = 1$: $\mathbf{N'} = \mathbf{L'}_1$,

i.e. $[N_x, N_y, N_z] + [n_x, n_y, n_z] = [L_{x1}, L_{y1}, L_{z1}] + t\,[v_{x1}, v_{y1}, v_{z1}]$ (6.22)

At $r = 0$: $\mathbf{N}' = \mathbf{L}'_2$,

i.e. $$[N_x, N_y, N_z] = [L_{x2}, L_{y2}, L_{z2}] + s\,[v_{x2}, v_{y2}, v_{z2}] \quad (6.23)$$

Subtracting Equation (6.23) from Equation (6.22) gives:

$$[n_x, n_y, n_z] =$$
$$[L_{x1} - L_{x2}, L_{y1} - L_{y2}, L_{z1} - L_{z2}] + [t\,v_{x1} - s\,v_{x2}, t\,v_{y1} - s\,v_{y2}, t\,v_{z1} - s\,v_{z2}] \quad (6.24)$$

The components of Equation (6.24) give the following three equations:

$$n_x - t\,v_{x1} + s\,v_{x2} = L_{x1} - L_{x2} \quad (6.25)$$
$$n_y - t\,v_{y1} + s\,v_{y2} = L_{y1} - L_{y2} \quad (6.26)$$
$$n_z - t\,v_{z1} + s\,v_{z2} = L_{z1} - L_{z2} \quad (6.27)$$

Both of the lines are perpendicular to the normal, therefore:

$$\mathbf{n} \,.\, \mathbf{v}_1 = n_x\,v_{x1} + n_y\,v_{y1} + n_z\,v_{z1} = 0 \quad (6.28)$$
$$\mathbf{n} \,.\, \mathbf{v}_2 = n_x\,v_{x2} + n_y\,v_{y2} + n_z\,v_{z2} = 0 \quad (6.29)$$

Equations (6.25 - 6.29) are solved to obtain s_m and t_m (the points where the two lines are closest to each other):

$$\begin{bmatrix} 1 & 0 & 0 & -v_{x1} & v_{x2} \\ 0 & 1 & 0 & -v_{y1} & v_{y2} \\ 0 & 0 & 1 & -v_{z1} & v_{z2} \\ v_{x1} & v_{y1} & v_{z1} & 0 & 0 \\ v_{x2} & v_{y2} & v_{z2} & 0 & 0 \end{bmatrix} \begin{bmatrix} n_x \\ n_y \\ n_z \\ t \\ s \end{bmatrix} = \begin{bmatrix} L_{x1} - L_{x2} \\ L_{y1} - L_{y2} \\ L_{z1} - L_{z2} \\ 0 \\ 0 \end{bmatrix} \quad (6.30)$$

Therefore: $$t_m = \frac{1}{\Delta}(-\delta\gamma + \omega\beta)\text{ , and} \quad (6.31)$$

$$s_m = \frac{1}{\Delta}(-\delta\beta + \omega\alpha) \tag{6.32}$$

Note that if D = 0 then there are an infinite number of solutions (i.e. the two line segments are parallel).

$$\begin{bmatrix} t \\ s \end{bmatrix} = \frac{1}{\Delta}\begin{bmatrix} \gamma & \beta \\ \beta & \alpha \end{bmatrix}\begin{bmatrix} -\delta \\ \omega \end{bmatrix} \Rightarrow \begin{bmatrix} -\delta \\ \omega \end{bmatrix} = \begin{bmatrix} \alpha & -\beta \\ -\beta & \gamma \end{bmatrix}\begin{bmatrix} t \\ s \end{bmatrix} \tag{6.33}$$

Therefore: $\alpha t - \beta s = -\delta$, and (6.34)

$$-\beta t + \gamma s = \omega \tag{6.35}$$

6.3.1.5 Two parallel Lines

If two lines are parallel, then the above method (Section 6.3.1.2) yields an infinite number of solutions, so create one independent variable instead of having two. Since $\mathbf{v}_1 + q\,\mathbf{v}_2 = \mathbf{0}$, set $w = q\,t + s$. To obtain the minimum distance between two parallel lines, select a fixed point on one line by setting the parametric variable to a constant (on line one set $t = 0$) and use an arbitrary point on the other line by leaving the parametric variable varying (on line two set $s = w$). This will give the same result as in Section (6.3.1.1) (refer to Figure 6.6).

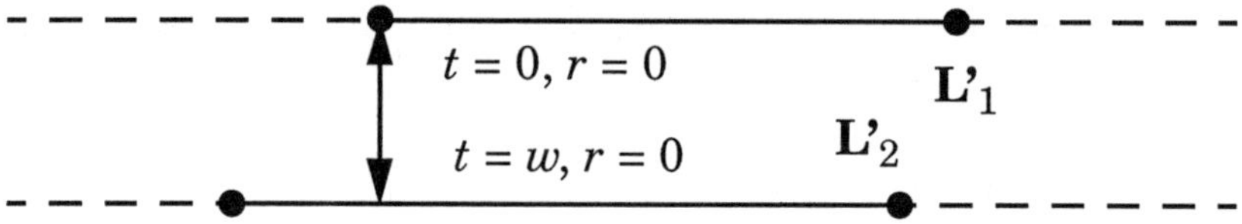

Figure 6.6: Two parallel lines

6.3.2 Minimum Distance Between Two Line Segments

The previous section (Section 6.3.1) needs to be modified for line segments - the manipulator links have finite lengths.

6.3.2.1 Two Non-Parallel Line Segments

For line segments both s and t must lie in the interval [0,1]. If both s and t are outside the interval [0,1] then compute the minimum distance from each of the end-points of the two lines to the other line segment (refer to Section 6.3.1.2) and select the minimum, i.e.

Let: d_1 = distance from point $\mathbf{L}_1$ on line $\mathbf{L'}_1$ to line segment $\mathbf{L'}_2$,
i.e. $t = 0$ and $s = w/g$ (Equation (6.35) with $t = 0$).

d_2 = distance from point $\mathbf{L}_1 + \mathbf{v}_1$ on line $\mathbf{L'}_1$ to line segment $\mathbf{L'}_2$,
i.e. $t = 1$ and $s = (w + b)/g$ (Equation (6.35) with $t = 1$).

d_3 = distance from point $\mathbf{L}_2$ on line $\mathbf{L'}_2$ to line segment $\mathbf{L'}_1$,
i.e. $s = 0$ and $t = -d/a$ (Equation (6.34) with $s = 0$).

d_4 = distance from point $\mathbf{L}_2 + \mathbf{v}_2$ on line $\mathbf{L'}_2$ to line segment $\mathbf{L'}_1$.
i.e. $s = 1$ and $t = (-d + b)/a$ (Equation (6.34) with $s = 1$).

Therefore: $d_{min} = \min \{d_1, d_2, d_3, d_4\}$

If either s or t, but not both, are outside the interval [0,1] then the variable outside must be reset, and the other needs to be recomputed using the minimum distance between a point (on the line where the variable was outside the interval at the closest end-point) and a line segment (the other line). Which results in:

If $t < 0$ then $t = 0$ and $s = w/g$ (Equation (6.35) with $t = 0$).
If $t > 1$ then $t = 1$ and $s = (w + b)/g$ (Equation (6.35) with $t = 1$).
If $s < 0$ then $s = 0$ and $t = -d/a$ (Equation (6.34) with $s = 0$).
If $s > 1$ then $s = 1$ and $t = (-d + b)/a$ (Equation (6.34) with $s = 1$).

6.3.2.2 Two Parallel Line Segments

For two parallel line segments, compute the minimum distance from each of the end-points of the two lines to the other line segment (refer to Section 6.3.1.2). The lesser of the four distances is the minimum distance (refer to Figure 6.7).

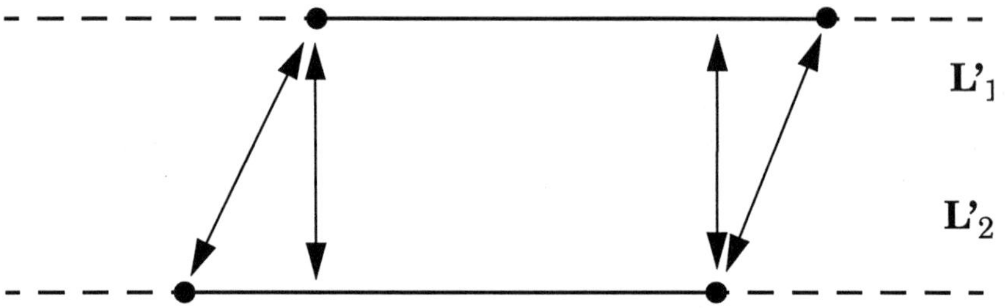

Figure 6.7: Two parallel lines

Use: d_1 = distance from point $\mathbf{L}_1$ on line $\mathbf{L'}_1$ to line segment $\mathbf{L'}_2$,
i.e. $t = 0$ and $s = w/g$ (Equation (6.35) with $t = 0$).

d_2 = distance from point $\mathbf{L}_1 + \mathbf{v}_1$ on line $\mathbf{L'}_1$ to line segment $\mathbf{L'}_2$,
i.e. $t = 1$ and $s = (w + b)/g$ (Equation (6.35) with $t = 1$).

d_3 = distance from point $\mathbf{L}_2$ on line $\mathbf{L'}_2$ to line segment $\mathbf{L'}_1$,
i.e. $s = 0$ and $t = -d/a$ (Equation (6.34) with $s = 0$).

d_4 = distance from point $\mathbf{L}_2 + \mathbf{v}_2$ on line $\mathbf{L'}_2$ to line segment $\mathbf{L'}_1$.
i.e. $s = 1$ and $t = (-d + b)/a$ (Equation (6.34) with $s = 1$).

Therefore: $d_{min} = \min \{d_1, d_2, d_3, d_4\}$

It may be better, in terms of the calculation load, to compute the minimum distance between two parallel lines (as in Section 6.3.2.1), and only if this distance is less than the threshold then compute the minimum distance between two parallel lines segments as above.

6.3.3 Definitions and Preliminary Calculations for Planes

Equation of a line: $\mathbf{L'} = \mathbf{L} + t\,\mathbf{v} = [L_x, L_y, L_z] + t\,[v_x, v_y, v_z]$

Arbitrary point on the line: $\mathbf{h} = [h_x, h_y, h_z]$

Equation of a plane: $a\,x + b\,y + c\,z + d = 0$

Arbitrary point on the plane: $\mathbf{k} = [k_x, k_y, k_z]$
Line connecting points **h** and **k**: $\overline{\mathbf{hk}}$

Normal (to plane): $\mathbf{n} = [a, b, c]$

Unit normal: $\hat{\mathbf{n}} = \mathbf{n}/|\mathbf{n}|$ (6.36)

Note: $$\cos\Theta = \frac{\overline{\mathbf{hk}} \cdot \mathbf{n}}{\|\overline{\mathbf{hk}}\|\,\|\mathbf{n}\|} = \frac{\overline{\mathbf{hk}} \cdot \hat{\mathbf{n}}}{\|\overline{\mathbf{hk}}\|} \tag{6.37}$$

Where Θ is the angle between line $\overline{\mathbf{hk}}$ and $\hat{\mathbf{n}}$.

The "Jordan Curve Theorem" (Watkins and Sharp 1992) can be used to detect if a point lies within any plane segment. It basically states that if a line is drawn away from the point to infinity in any direction, and the number of intersections it has with any of the boundary line segments is odd then the point lies within the plane segment, otherwise if it is even then the point lies outside the plane segment. Rather than use infinity, use an infinity-approximation, i.e. a very large coordinate or double the largest co-ordinate of the plane segment. In Figure (6.8), the line from point A crosses three boundary line segments therefore it lies inside the plane segment, and the line from point B crosses two boundary line segments therefore it lies outside the plane segment.

A line, $\mathbf{L}' = \mathbf{L} + t\,\mathbf{v}$, is parallel to a plane, $a\,x + b\,y + c\,x + d = 0$, if:

$$v \cdot [a, b, c] = 0 \tag{6.38}$$

i.e. $$av_x + bv_y + cv_z = 0 \tag{6.39}$$

6.3.3.1 Plane and a Parallel Line (Segment)

The distance from a parallel line (or parallel line segment) to the plane (not a plane segment), as shown in Figure (6.9):

$$d = \left| \|\overline{\mathbf{hk}}\| \cos\Theta \right| = \left| \overline{\mathbf{hk}} \cdot \hat{\mathbf{n}} \right| \tag{6.40}$$

$$= \left| \frac{\left(h_x - k_x, h_y - k_y, h_z - k_z\right) \cdot (a, b, c)}{\sqrt{a^2 + b^2 + c^2}} \right| \tag{6.41}$$

$$= \frac{\sqrt{a^2\left(h_x - k_x\right)^2 + b^2\left(h_y - k_y\right)^2 + c^2\left(h_z - k_z\right)^2}}{\sqrt{a^2 + b^2 + c^2}} \tag{6.42}$$

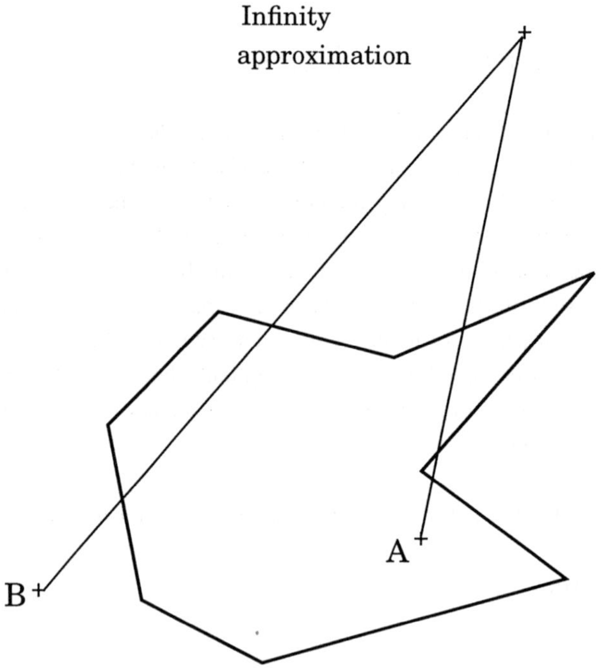

Figure 6.8: Point within a plane segment

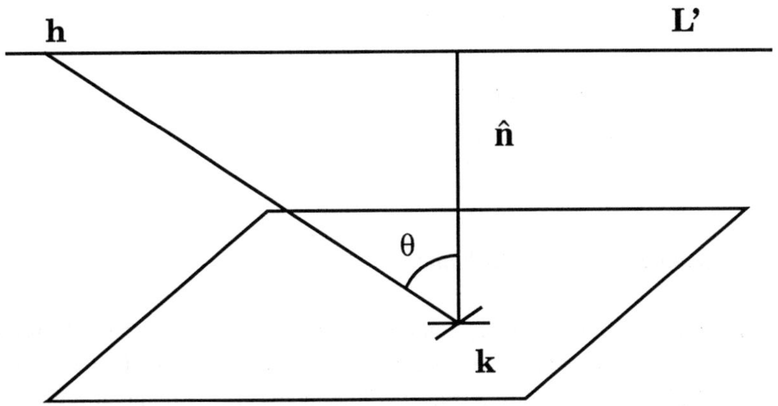

Figure 6.9: Distance between a plane segment and a parallel line

6.3.3.2 Line and Plane Intersection

If a line (**L'** = **L** + t**v**) is not parallel with a plane ($ax + by + cz + d = 0$) then the line will intersect the plane at some point (at t_i on the line), as shown in Figure (6.10):

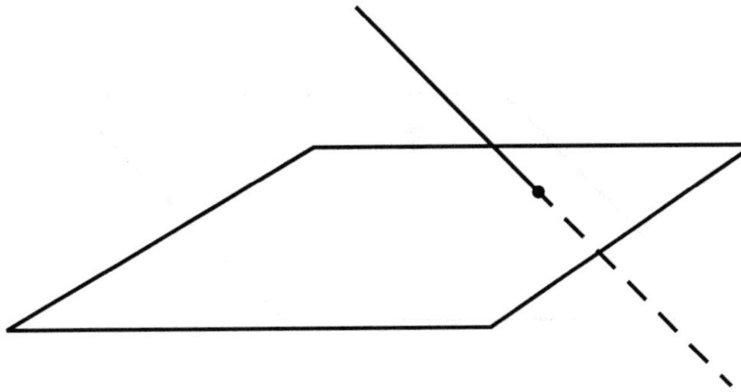

Figure 6.10: Line intersecting a plane

$$a\,(L_x + t_i\,v_x) + b\,(L_y + t_i\,v_y) + c\,(L_z + t_i\,v_z) + d = 0 \tag{6.43}$$

i.e.
$$t_i = -\frac{aL_x + bL_y + cL_z + d}{av_x + bv_y + cv_z} \tag{6.44}$$

6.3.3.3 Projecting a Point onto a Plane

When projecting a point $[x_p, y_p, z_p]$ onto the plane $a\,x + b\,y + c\,z + d = 0$ (refer to Figure 6.11), a line formed from the point to the projected point is: $\mathbf{L}' = \mathbf{L} + t\,\mathbf{v}$. Set $\mathbf{L} = [x_p, y_p, z_p]$ and $\mathbf{v} = \mathbf{n} = [a, b, c]$. This line intersects the plane at point $t = t_i$, where:

$$a\,(x_p + t_i\,a) + b\,(y_p + t_i\,b) + c\,(z_p + t_i\,c) + d = 0 \tag{6.45}$$

i.e.
$$t_i = -\frac{ax_p + by_p + cz_p + d}{a^2 + b^2 + c^2} \tag{6.46}$$

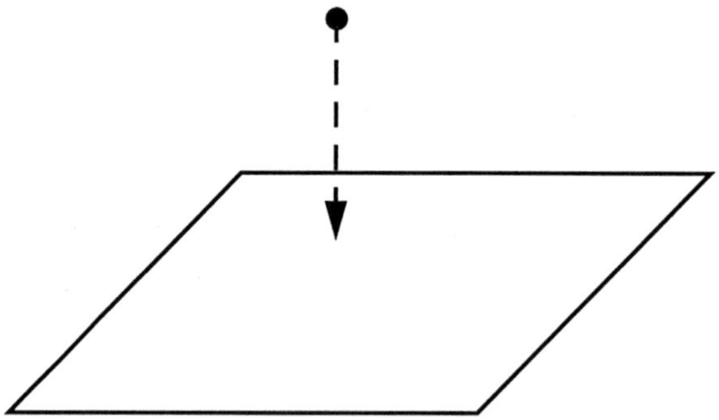

Figure 6.11: Projecting a point onto plane

The distance from the point to the plane is:

$$d = t_i \sqrt{a^2 + b^2 + c^2} \tag{6.47}$$

6.4 Algorithm

The following algorithm implements the routine to compute the minimum distance from a line segment to a plane segment.

1. **If** the line and plane are parallel **then goto step** 2 **else goto step** 8.

2. Compute the distances from each boundary line of the plane segment to the line segment (refer to Figure 6.12):

d_i = the distance from boundary line i to the line segment.

Where: $i = 1, 2, ..., m$.

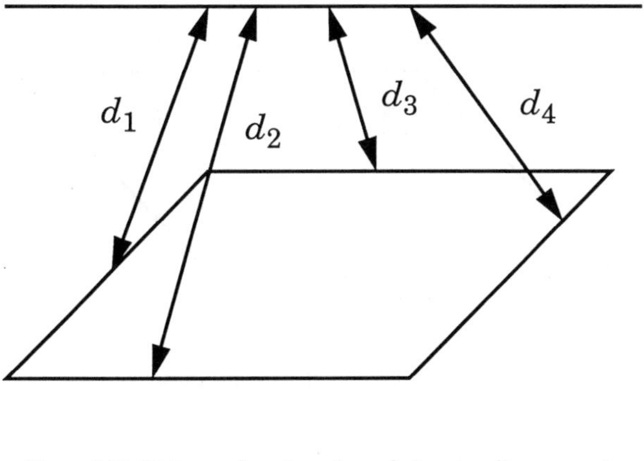

Figure 6.12: Distances from boundary of plane to a line segment

3. Project the end-points of the line segment onto the plane (refer to Figure 6.13).

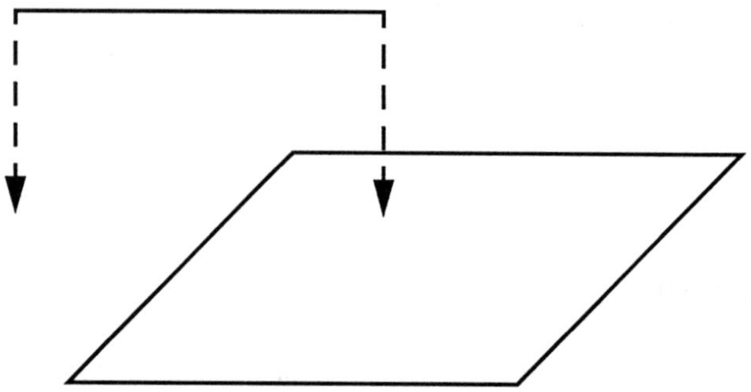

Figure 6.13: Projection of line segment end-points onto a plane

4. **If** both projected end-points are in the plane segment **then goto step** 5 **else goto step** 7.

5. Compute the projected length from one end-point, denoted *d* (refer to Figure 6.14).

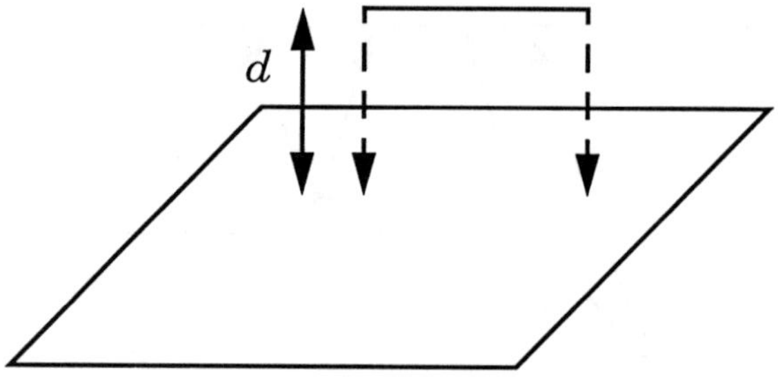

Figure 6.14: Projected length distance

6. **Return** *d* as the minimum distance **and stop**.

7. **Return** $d = \min\ \{d_1, d_2, ..., d_m\}$ as the minimum distance **and stop**.

8. Compute the intersection point between the line and the plane (refer to Figure 6.10).

9. **If** the intersection point lies in both the line segment and the plane segment **then goto step** 10 **else goto step** 11.

10. **Return** 0 as the minimum distance **and stop.**

11. **Perform step** 2.

12. **Perform step** 3.

13. **If** either of the end points lie inside the plane segment **then goto step** 14 **else goto step** 16.

14. Compute the projection lengths (d_a and d_b (if it exists)), as in Figures (6.15) and (6.16).

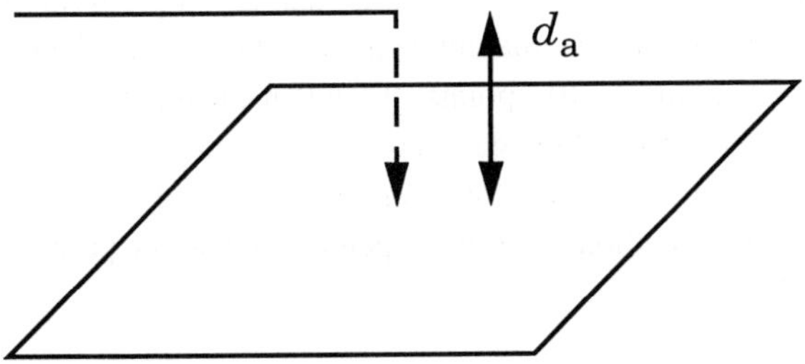

Figure 6.15: Single projection length

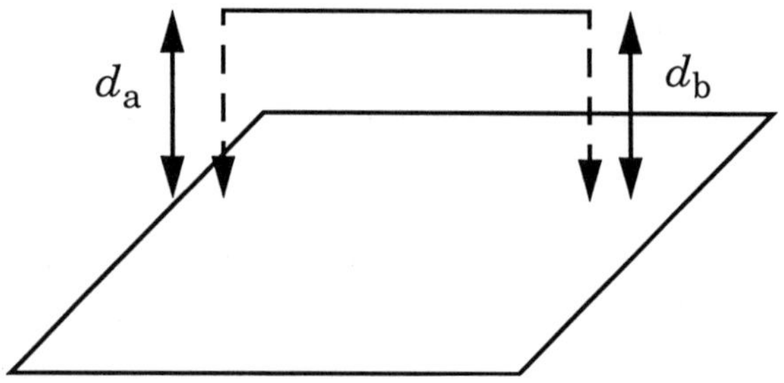

Figure 6.16: Twin projection lengths

15. **Return** $d = \min \{d_1, d_2, ..., d_4, d_a(, d_b)\}$ as the minimum distance **and stop**.

16. **Return** $d = \min \{d_1, d_2, ..., d_m\}$ as the minimum distance **and stop.**

6.5 Results and Examples

Two examples are presented. In the first, a series of snapshots, Figures (6.17) - (6.23), show two 3 DOF manipulators and a single object (a rectangular plane) in the same workspace. They are taken from an animation program and demonstrate collisions occurring among the manipulators and the object. The manipulators follow a trajectory containing 101 points, the left-hand image is a top view of the workspace and the right-hand image is a rotated view of the workspace. Table (6.1) contains the links parameters for the manipulators. The trajectories are generated using the polynomial method (refer to Appendix 2), and the parameters for the trajectories are in Table (6.2).

In the second example, a series of snapshots, Figures (6.24) - (6.29), show a two DOF planar manipulator and a single object (a rectangular plane) in the same workspace. The manipulator follows a trajectory containing 101 points, the image is a top view of the workspace. Table (6.3) contains the links parameters for the manipulators. The trajectories are generated using the polynomial method (refer to Appendix 2), and the parameters for the trajectories are in Table (6.4).

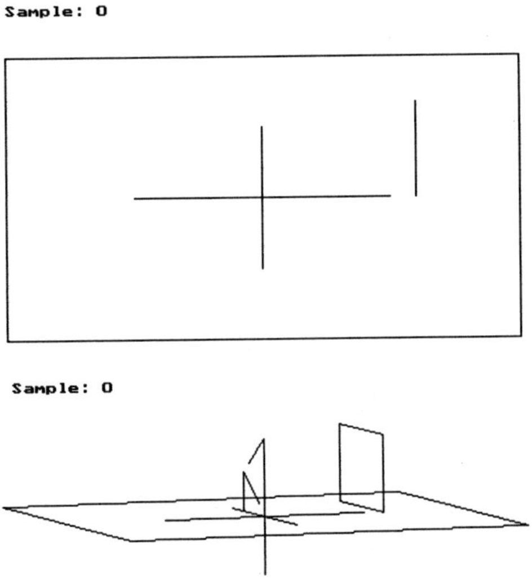

Figure 6.17: Sample 0 - no collisions

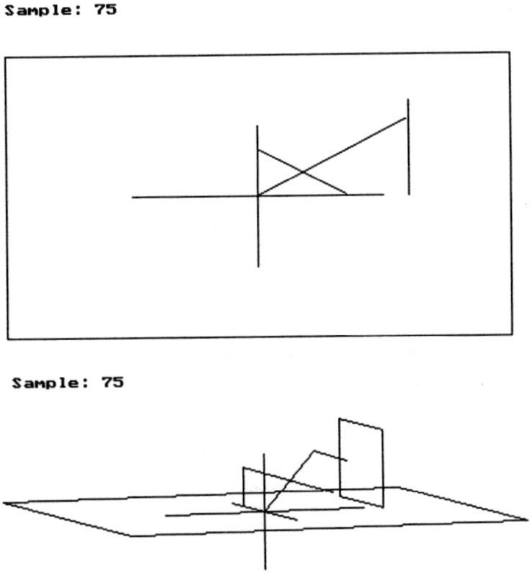

Figure 6.18: Sample 75 - no collisions

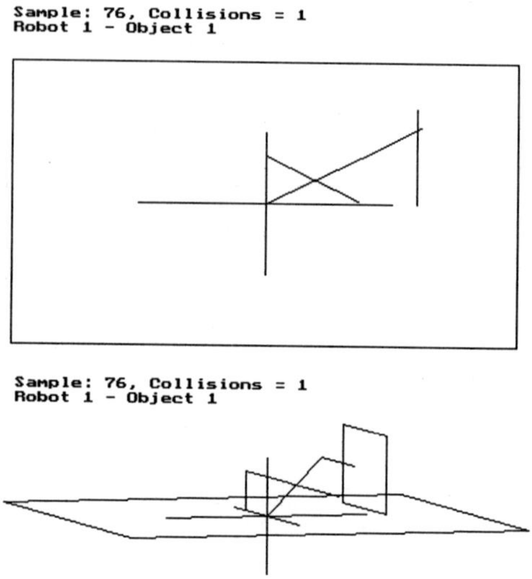

Figure 6.19: Sample 76 - one collision

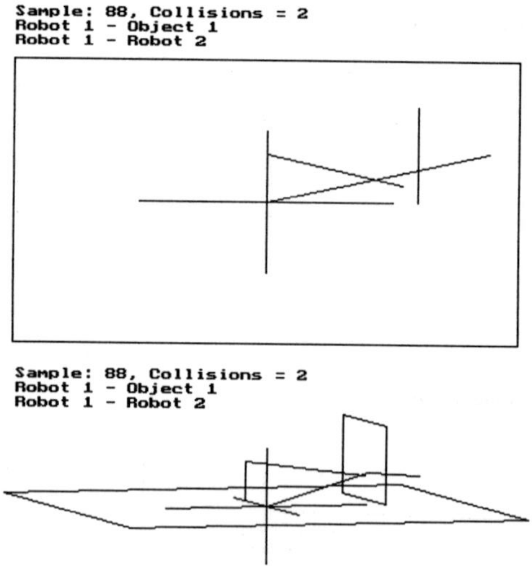

Figure 6.20: Sample 88 - two collisions

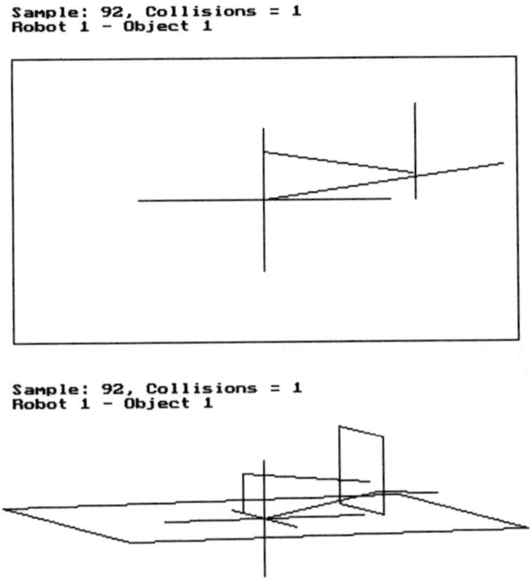

Figure 6.21: Sample 92 - one collision

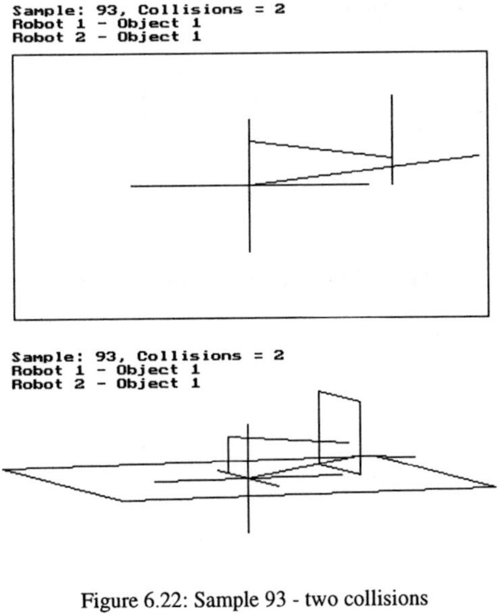

Figure 6.22: Sample 93 - two collisions

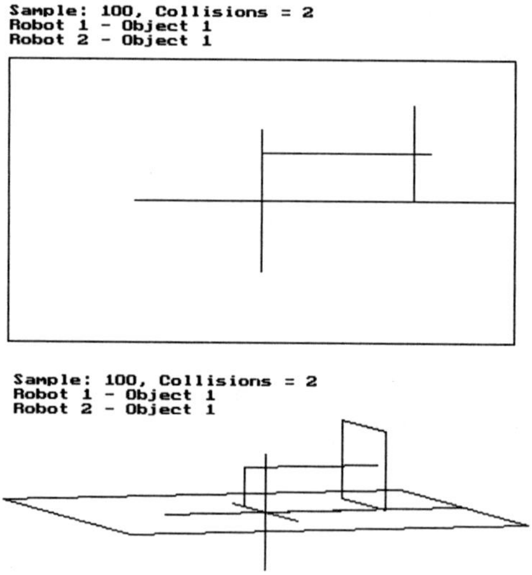

Figure 6.23: Sample 100 - two collisions

Robot	Link	Link type	a_i (m)	α_i (rad)	d_i (m)	θ_i (rad)	Link thickness (m)	$q_{\min}$	$q_{\max}$
1	1	Revolute	0	$\pi/2$	0	Variable	0.01	0	2π
1	2	Revolute	2	0	0	Variable	0.01	0	2π
1	3	Revolute	1	0	0	Variable	0.01	0	2π
2	1	Revolute	0	$-\pi/2$	1	Variable	0.01	0	2π
2	2	Revolute	0	$\pi/2$	0	Variable	0.01	0	2π
2	3	Prismatic	0	0	Variable	0	0.01	0	2π

Table 6.1: Link parameters

Robot	Link	q_i	q_f	q'_i	q'_f	q''_i	q''_f
1	1	$\pi/2$	0	0	$-\pi/10$	0	0
1	2	$\pi/2$	0	0	$-\pi/10$	0	0
1	3	$3\pi/4$	0	0	$3\pi/20$	0	0
2	1	$-\pi/2$	0	0	$\pi/10$	0	0
2	2	$3\pi/4$	$\pi/2$	0	$-\pi/10$	0	0
2	3	1	2	0	0.2	0	0

Table 6.2: Trajectory parameters

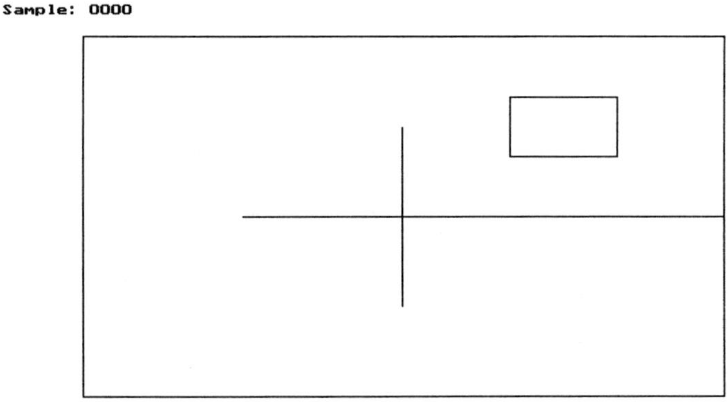

Figure 6.24: Sample 0 - no collisions

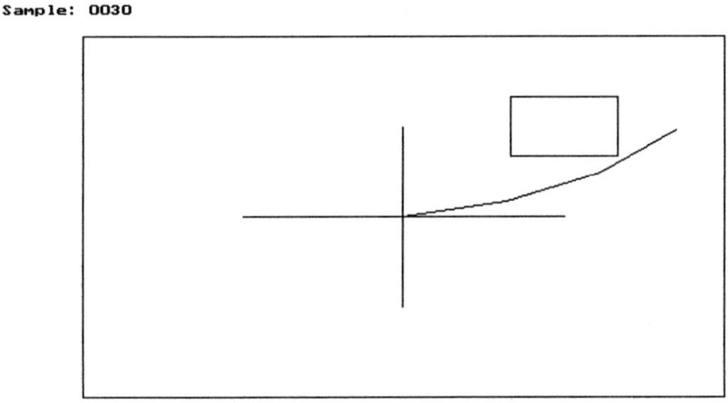

Figure 6.25: Sample 30 - no collisions

Sample: 0031, Collisions = 01.
Robot 01 - Object 01

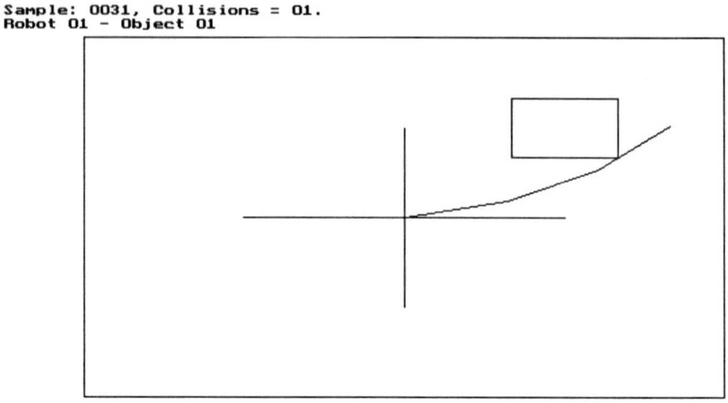

Figure 6.26: Sample 31 - one collision

Sample: 0046, Collisions = 01.
Robot 01 - Object 01

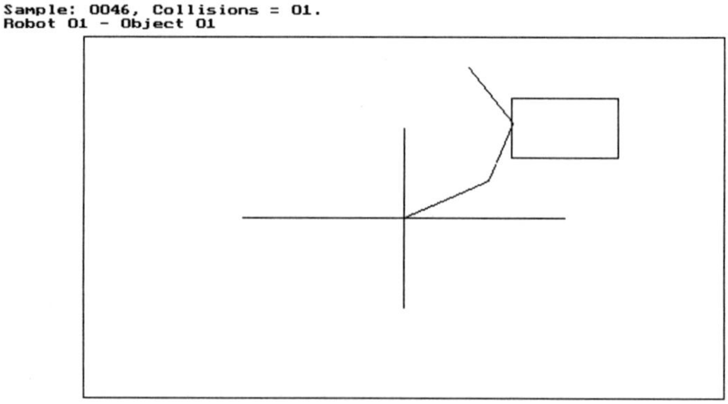

Figure 6.27: Sample 46 - one collision

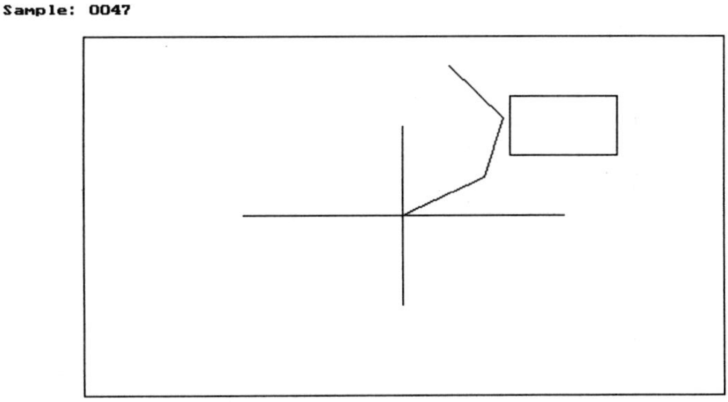

Figure 6.28: Sample 47 - no collisions

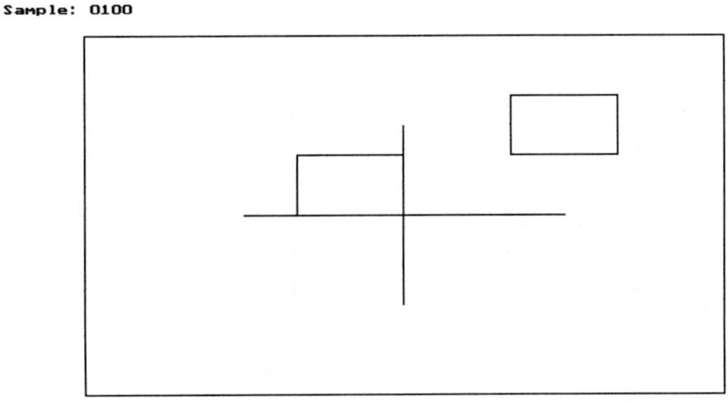

Figure 6.29: Sample 100 - no collisions

Link	Link type	a_i (m)	α_i (rad)	d_i (m)	θ_i (rad)	Link thickness (m)	$q_{\min}$	$q_{\max}$
1	Revolute	1	0	0	Variable	0.01	0	2π
2	Revolute	1	0	0	Variable	0.01	0	2π
3	Revolute	1	0	0	Variable	0.01	0	2π

Table 6.3: Link parameters

Link	q_i	q_f	q'_i	q'_f	q''_i	q''_f
1	0	$\pi/2$	0	0	0	0
2	0	$\pi/2$	0	0	0	0
3	0	$\pi/2$	0	0	0	0

Table 6.4: Trajectory parameters

6.6 Conclusions

This chapter has demonstrated a technique to detect collisions using approximate representations of manipulators and objects. The accuracy depends on how close the robot manipulator model is to the real robot manipulator, how the radius of each link is chosen (an average value or the maximum possible value), and how close the object model is to the real object. The object descriptions are constrained by being only made up of planes and spheres. Most of the methods used can take advantage of parallel processing (Zomaya 1992) to speed up the calculations, especially where a large number of links and/or plane segments are involved. The examples used demonstrated the algorithm successfully operating. The algorithm will be used in the next chapter for collision avoidance to compute distances between manipulators and obstacles.

CHAPTER 7

COLLISION AVOIDANCE

7.1 Introduction

The path planning problem for manipulators asks to find collision free paths for manipulators moving in an environment populated with obstacles (some of these obstacles can be other manipulators). The method presented here is straightforward and generates collision free trajectories for several manipulators working in the same environment. The end-effector of each manipulator can move from a specified starting point to a particular goal without colliding with any other manipulators or objects in the three dimensional environment (the objects can either be moving, stationary, or shape changing). There are two methods used to guide the end-effector in this work: (1) Potential fields; and (2) Cell decomposition. GAs are used to compute the required joint angles. The level of difficulty in solving this problem depends on the construction of the manipulator(s) and the structure of the environment.

Various methods have been used in the literature to coordinate robot manipulators along collision free trajectories. Most of these methods use configuration space (a transformation from cartesian space (Latombe 1991)), where the manipulator is treated as a point in configuration space (and everything else in the environment is transformed as well). This point is guided through the environment. Usually this guidance is provided by potential fields which exert an attractive force towards the goal and a repulsive force away from obstacles. Examples of this can be found in (Latombe 1991; Bessiere 1993; Solano and Jones 1993). However not all methods use configuration space, for example, the work by Khoogar and Parker (1991) encodes the complete set of joint motions for a manipulator in a genetic string for processing by a GA.

The approach proposed here uses cartesian space, rather than configuration space (Gill and Zomaya 1996; Gill and Zomaya 1998a; Gill and Zomaya 1998b). Our approach is simpler and no complex time consuming mapping of the whole workspace needs to be performed. Other drawbacks of the configuration space approach are that the initial and final configurations must be known, the dimension of the space is equal to the DOF of the manipulator (i.e. as the manipulator complexity grows then so does the dimension of the configuration space), and a complete description of the environment must be known in advance. It would be reasonable to expect the initial configuration to be known, but the final may not be known, if it is desired that the end-effector be moved to a particular position without regard to how the links are positioned.

The end-effector of each manipulator is guided iteratively from its starting position towards its goal and away from obstacles using potential fields or graph searching in approximate cell decomposition. It does not require complete knowledge of the workspace - only local knowledge is used. The local knowledge required is the distance from the end-effector to the nearest obstacles and the minimum distance from the whole manipulator to the closest obstacles. The GA is used to find suitable joint variables for each manipulator to ensure that no collision occurs with other obstacles (i.e. objects and other manipulators) and that the end-effector moves close enough to the desired target during each iteration.

7.2 Particle Motion Through Space

The guidance of manipulators through cartesian space involves guiding the end-effector (which is the same as moving a particle through space) and then calculating the required joint variables to realize this position without any part of the manipulator colliding with any obstacles. Therefore the first part of the problem is to move a particle through space without it touching any obstacles. The two methods used are presented in the next two sections (Section (7.2.1) and Section (7.2.2)).

7.2.1 Potential Field

The potential field approach creates an artificial potential field (U) in space which reflects the structure of space. It consists of two components: (1) An attractive potential (U_{att}) pulling a particle towards the goal which monotonically increases as the distance from the goal increases; and (2) A repulsive potential (U_{rep}) pushing the particle away from obstacles, which is generated at the position of the obstacles and has a high value at this position which decreases monotonically as the distance from the obstacle increases. The overall potential is the superposition of the attractive and repulsive components. A particle in space, at point $\mathbf{q}$, will be under the influence of U. The artificial force acting on it is the negative gradient of U, which is (Latombe 1991):

$$\mathbf{F}(\mathbf{q}) = \mathbf{F}_{att}(\mathbf{q}) + \mathbf{F}_{rep}(\mathbf{q}) = -\nabla U(\mathbf{q}) \tag{7.1}$$

The attractive potential is felt over the whole region and is defined as:

$$U_{att}(\mathbf{q}) = \frac{1}{2}\varepsilon\rho_{goal}^{2}(\mathbf{q}) \tag{7.2}$$

Where the Euclidean distance to the goal is:

$$\rho_{goal}(\mathbf{q}) = \|\mathbf{q} - \mathbf{q}_{goal}\| \tag{7.3}$$

Therefore the attractive force is:

$$\mathbf{F}_{att} = -U_{att}(\mathbf{q}) = -\varepsilon\left(\mathbf{q} - \mathbf{q}_{goal}\right) \tag{7.4}$$

The repulsive potential away from an obstacle is only felt in the vicinity of obstacles (i.e. within a distance r_0) and is defined as:

$$U_{rep}(\mathbf{q}) = \begin{cases} \frac{1}{2}\eta\left(\frac{1}{p(\mathbf{q})} - \frac{1}{\rho_0}\right), p(q) \le \rho_0 \\ 0, p(q) > \rho_0 \end{cases} \tag{7.5}$$

Where the distance to the obstacle is:

$$p(\mathbf{q}) = \min_{\mathbf{q}' \in \text{object}} \|\mathbf{q} - \mathbf{q}'\| \tag{7.6}$$

Therefore the repulsive force away from an obstacle is:

$$\mathbf{F}_{rep}(\mathbf{q}) = \begin{cases} \eta\left(\frac{1}{p(\mathbf{q})} - \frac{1}{\rho_0}\right)\frac{1}{p^2(\mathbf{q})}\nabla p(\mathbf{q}), p(q) \le \rho_0 \\ 0, p(q) > \rho_0 \end{cases} \tag{7.7}$$

Where: e = Positive scaling factor,
$\mathbf{q}_{goal}$ = Location of the goal,

h = Positive scaling factor,
q' = Point on object closest to location **q**, and
r_0 = Distance of influence (positive).

The repulsive force from an obstacle is calculated if the particle, at **q**, is within its distance of influence and then the total repulsive force is the sum of each individual repulsive force. If the particle, at **q**, collides with the obstacle (i.e. the distance from the obstacle is zero) then the repulsive force from that obstacle will be infinite. The path is generated by following the negative gradient from the start to the goal. If the particle, at **q**, is to move a distance δd, then in the next iteration the new position will be:

$$\mathbf{q}^{t+1} = \mathbf{q}^{t} + \frac{\mathbf{F}}{\|\mathbf{F}\|} \times \delta d \tag{7.8}$$

In Equation (7.8) the force is normalized since only the direction is required. Figure (7.1) shows the shape of an attractive field potential, Figure (7.2) shows the shape of a repulsive field potential around an obstacle, and Figure (7.3) shows the overall potential (the sum of the repulsive and attractive potentials).

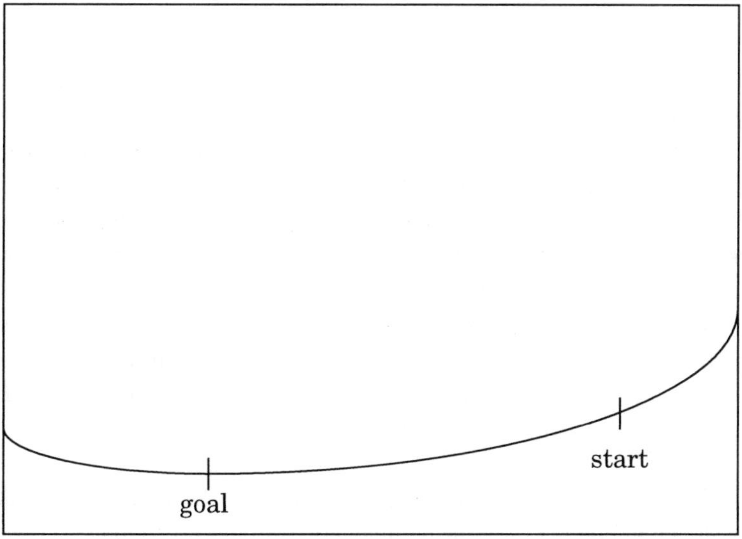

Figure 7.1: Attractive potential

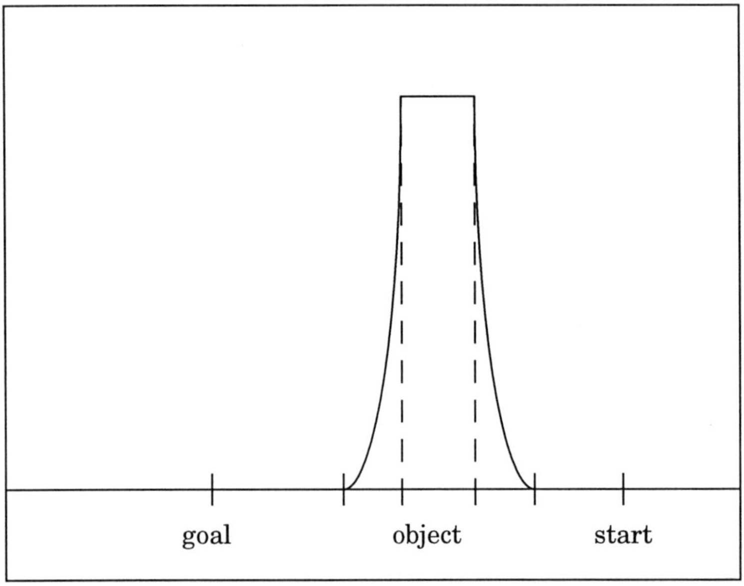

Figure 7.2: Repulsive potential

The problem with potential fields is that local minima exist, and at these points the net force is zero. So, it is possible to become trapped at a local minima. There exist many methods to handle this situation. In this work when a particle becomes trapped it will be temporarily diverted in another direction towards an alternate goal. To detect if the particle is staying in a particular region (where it is trapped) it will be in the vicinity of an obstacle (the distance from this obstacle (the closest obstacle) is known as the "vicinity distance") and it will not have moved very much (its distance from any previous locations is less then the "staying distance"). The particle will move towards the temporary goal (which should be suitably selected) until it is well out of the way of the obstacle (when it has move further than the "escape distance" from the obstacle).

A two dimensional example of a particle moving under the influence of a potential field is shown in Figure (7.4). The particle moves from [4, 1.5] to [1, 1.5] with the alternate goal at [2, 0]. The obstacle is a square plane, with vertices at coordinates:

[2, 1], [3, 1], [3, 2], and [2, 2]. Table (7.1) lists the potential field parameters used. It took 57 steps to reach the goal.

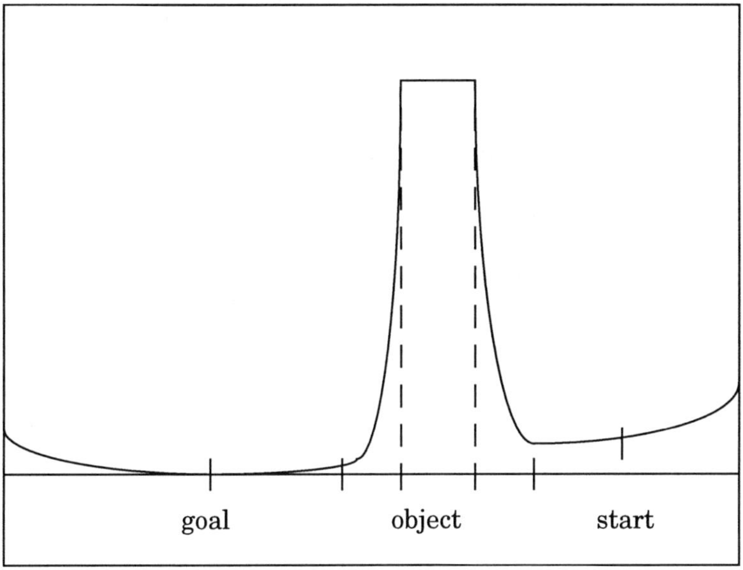

Figure 7.3: Overall potential

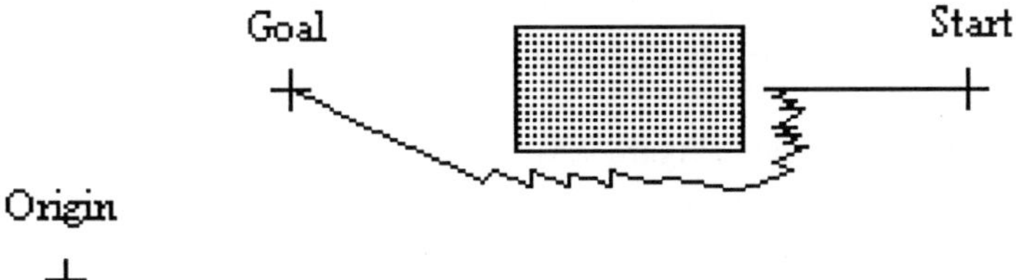

Figure 7.4: Potential field point motion

δ	0.1
ε	1
η	1
ρ	0.2
Staying distance	0.02
Escape distance	0.3
Vicinity distance	0.2

Table 7.1: Potential field parameters

7.2.2 Cell Decomposition

In cell decomposition (Latombe 1991) the entire workspace is divided into a number of non-overlapping cells and then the space is searched using a graph searching algorithm. The nodes of the graph represent each cell and the connections between them are to the adjacent cells. The graph is searched from the starting node to the end-point (goal). There are two types of cell decomposition: (1) Exact cell decomposition (object-dependent decomposition); and (2) Approximate cell decomposition (object-independent decomposition).

In exact cell decomposition (Figure 7.5) all the cells do not have to be the same size, but they contain only free space. The method also requires a complete description of the objects and a complex cell construction. After the cells are created a connectivity graph (Figure 7.6) is formed which represents the adjacency relationship between the cells. The connectivity graph is searched to find a solution - the cell path (Figure 7.7). From this a path is produced (Figure 7.8).

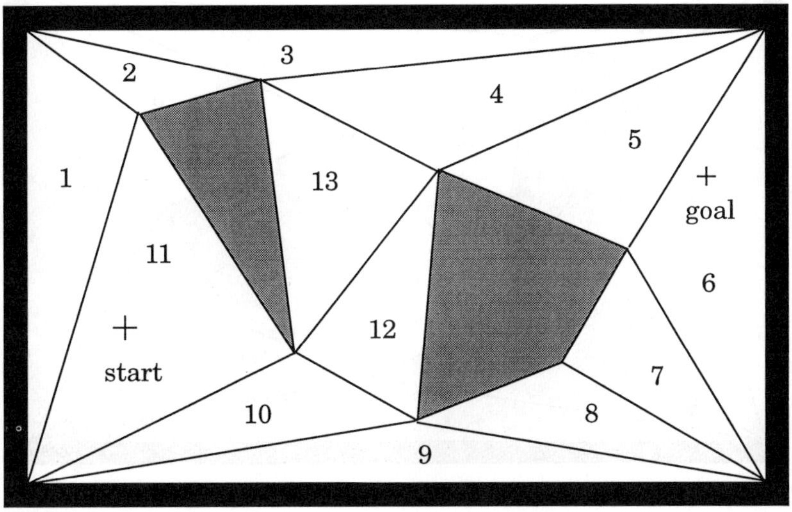

Figure 7.5: Exact cell decomposition

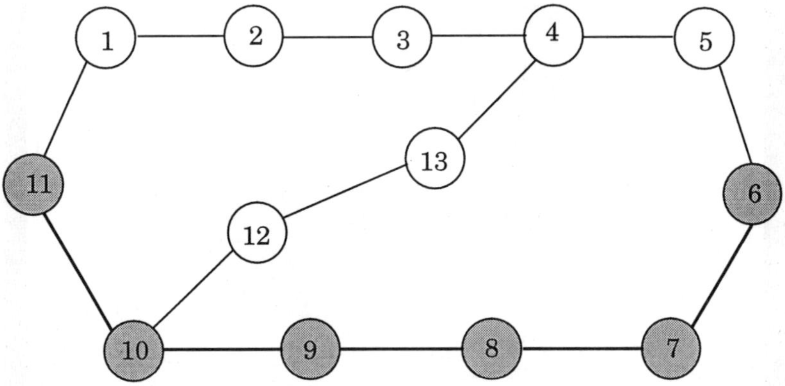

Figure 7.6: Connectivity graph

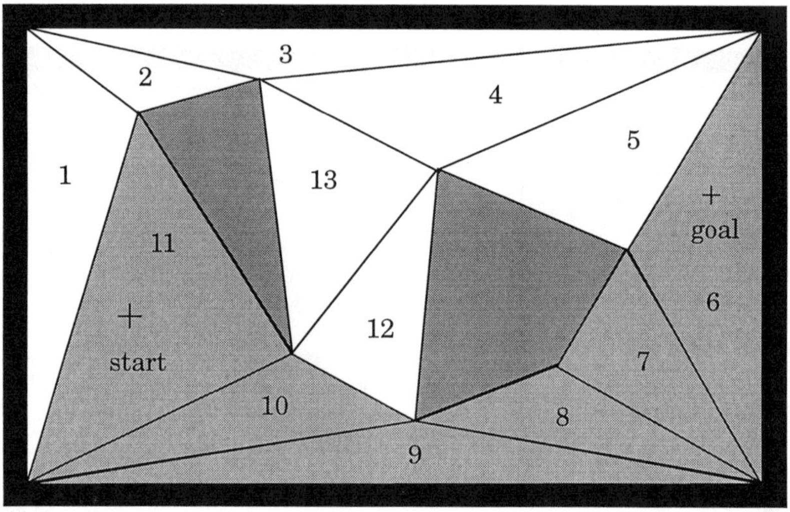

Figure 7.7: Cell path

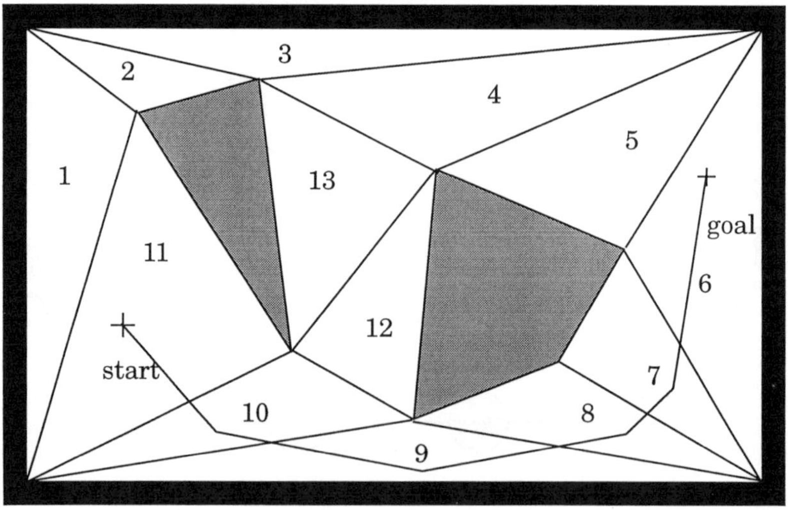

Figure 7.8: Final path

In approximate cell decomposition (Figure 7.9) the entire region is divided into equal sized cells. Any cell with any part of an obstacle in it is marked as invalid (full or occupied). This approach does not require a complete description of the workspace, as cells can be marked invalid as detected. The cells do not necessarily have to be marked as empty or full, but can be represented by a fraction of the portion occupied. The approximation can be improved by increasing the resolution but this will increase the memory requirements if the entire grid is stored in memory. The section of the grid currently in use can be in main memory or only the used parts of the grid can be stored in memory as a linked list or tree structure. The size of the cell used in this work is the distance that the end-effector can move during one time interval.

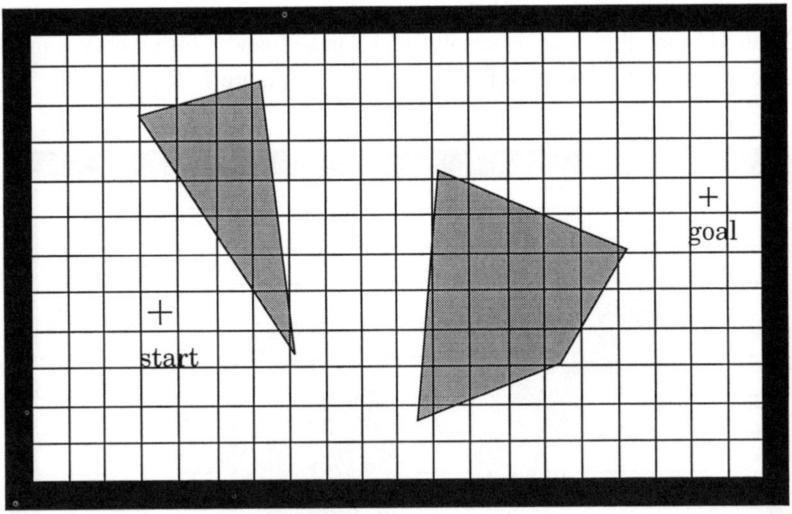

Figure 7.9: Approximate cell decomposition

This work uses the approximate cell decomposition method. The cells are cuboid and will be marked as empty or full. A partially filled cell is also marked as full. The workspace is initially divided up into cells. The cells are positioned such that the starting point is in the centre of a cell. The search starts from the cell containing the start and should conclude at the cell containing the goal. The cells are searched using a greedy depth-first search (Sedgewick 1990; Cormen *et al.* 1990). During each iteration the movement can only be to a neighbouring cell. Each cell visited is marked as visited. In two dimensions there are four neighbours - **top**, **left**, **bottom**, and **right**. In three dimensions there are six neighbours (only the neighbours along the x, y, and z axes are considered). The move is always to the next suitable cell that is closest to the goal.

Each neighbour is ranked according to its proximity to the goal (the closest being first). Then they are considered in turn (starting from the closest) to select the next cell to which to move. A cell is considered suitable if: (1) It does not exist outside the workspace; (2) If moving from the current cell the this cell does not cause a collision with an obstacle; and (3) It is not a previously visited cell (to avoid moving around in circles). If neither of the neighbouring cells are suitable, then the current cell is marked as visited and it backtracks to the previously occupied cell and the process repeats. If backtracking occurs all the way to the start and no further exploration is possible, then no suitable path can be found. This process of moving from cell to cell continues until the goal is reached (Figure 7.10).

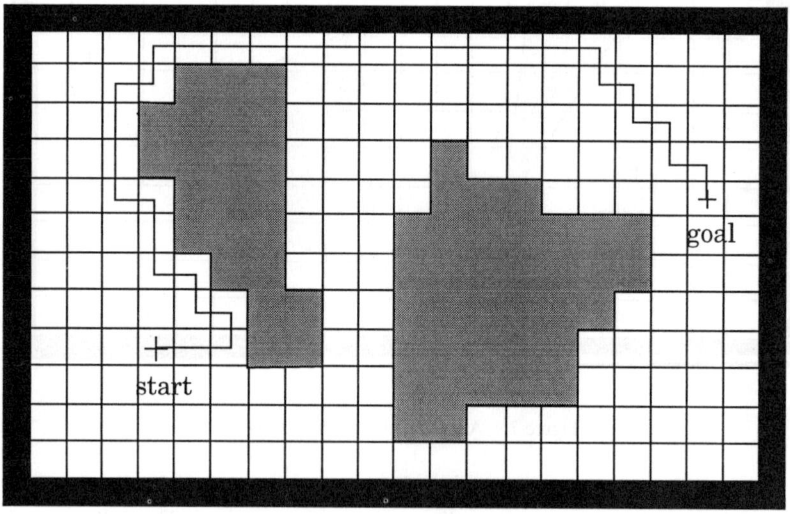

Figure 7.10: Final path

Figure (7.11) shows the options at the start in a two dimensional case. The movement can be to the top, left, bottom, or right. If the movement is to the right, then there will be only three possible moves. Movement back to the start in the left direction is impossible because the cell on the left has already been visited. The search can be represented as a graph as shown in Figure (7.12).

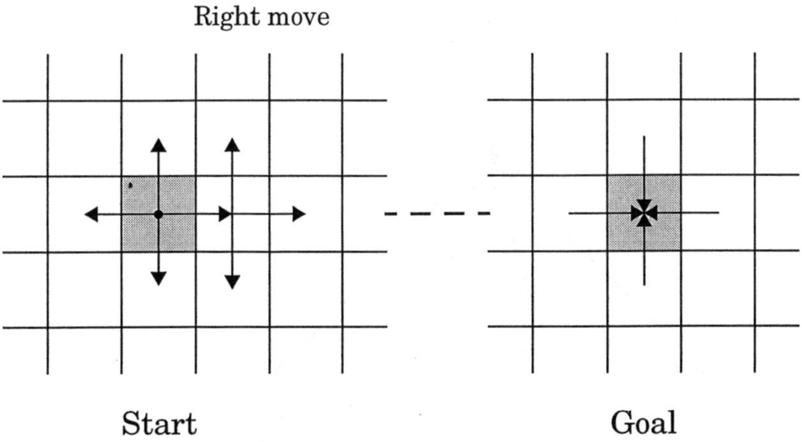

Figure 7.11: Cell neighbours

A two dimensional example of a particle moving through a region of space divided into cells using a greed depth first search is shown in Figure (7.13). The particle moves from [190, 40] to [10, 160] in steps of 10 units. The obstacle is a rectangle plane, with vertices at coordinates: [80, 20], [120, 20], [120, 100], and [80, 100]. Table (7.2) lists the parameters used. It took 31 steps to reach to goal.

7.3 Manipulator Implementation

The collision avoidance algorithm generates collision free trajectories for manipulators, with the end-effector of each moving from a specified starting point to a goal. To guide a manipulator, the end-effector is guided through the workspace using either the potential field (Section 7.2.1) or approximate cell decomposition (Section 7.2.2). The joint variables for the manipulator are calculated as in Section (4.2) with a modified fitness function. The GA searches for suitable joint angles for each manipulator, between allowable limits, to position the end-effector of each manipulator as close to the desired target as possible without any part of the manipulator colliding with any obstacle in the workspace. The fitness function for the GA contains an extra error term based on the proximity of the manipulator to obstacles. The manipulators move together towards their goals iteratively. During each iteration each manipulator is considered in turn, and each will have priority over the next (i.e. the first will have the highest priority).

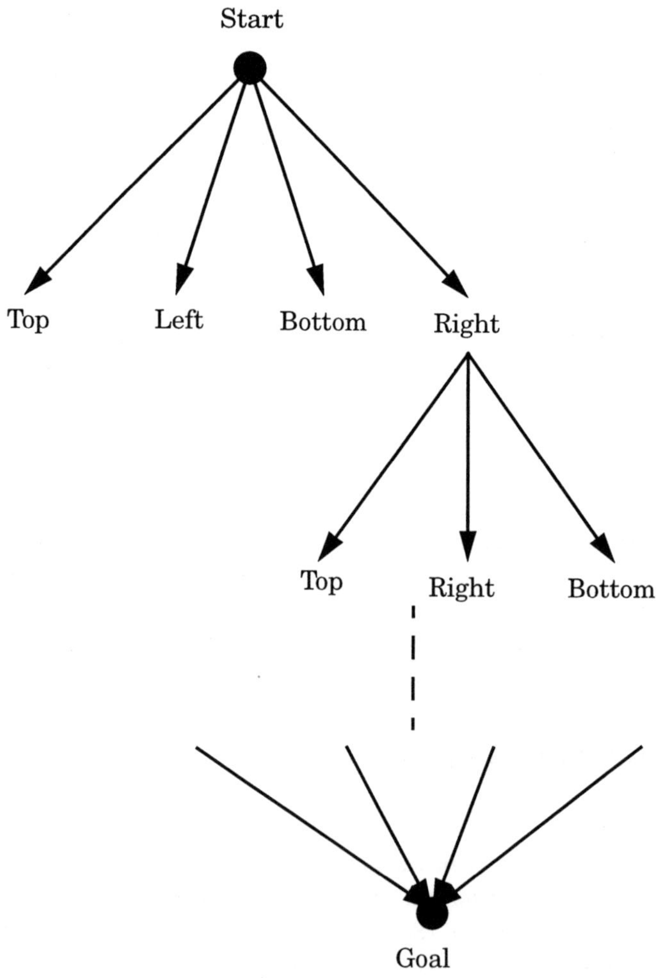

Figure 7.12: Search graph

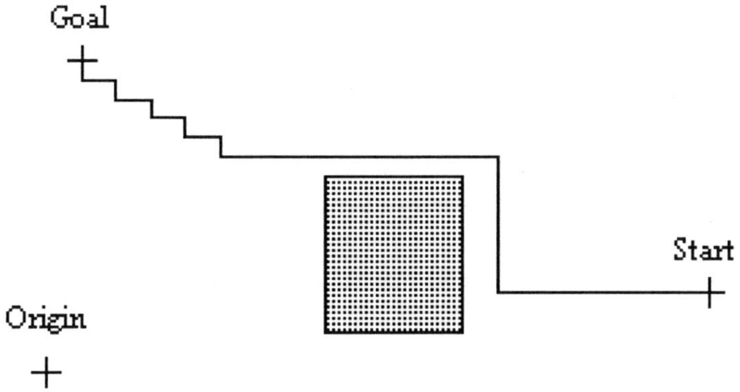

Figure 7.13: Cell decomposition point motion

Grid size	200 x 200
Cell size	10 x 10

Table 7.2: Cell decomposition parameters

The direction to the target is calculated from the current position of the end-effector using the potential field method or the cell decomposition method. The distance to be moved in each iteration is δd in time interval δt. In cell decomposition the cell size is δd. The GA is used in the same way as Section (4.2) to compute the required joint angles. If any of the manipulators in the environment reach their goals while other manipulators and objects are still moving then they may still need to have their joint variables computed, to move out of the way of other manipulators or objects if there is a chance of a collision. If a manipulator was unable to proceed then the procedure is aborted due to failure.

7.3.1 Potential Field Approach

If, during the motion using the potential field method, a manipulator end-effector is staying in the same region (this is caused by a local minima in the potential field or a moving obstacle pushing it around in circles) then it is temporarily moved in another direction (towards a temporary alternate goal) until it is out of the influence of this obstacle (i.e. it has moved a certain distance away from the obstacle). Then it will proceed as usual, in the same manner as in Section (7.2.1). The position of the alternate goal used here is the location of the base of the manipulator, this forces the manipulator to move around the near side of the obstacle rather than the far side where it may become trapped.

The above method, while considering the position of the end-effector, does not consider the speed and acceleration. If δd is always fixed over δt then the speed is always constant, but often a manipulator starts from a stationary position and finishes the same way. To account for the effects of varying speed the value of δd can be adjusted during the motion. An outline of the algorithm is shown in Figure (7.14).

There is a case, see Figure (7.15), where the manipulator becomes trapped. It occurs when the end-effector is free to move, but the proximal links of the manipulator are being blocked by an obstacle close to the base of the manipulator. One way to avoid this situation is to detect its occurrence and then force the manipulator to backtrack beyond the obstacle and then move around it in another direction. A simpler method would be to set an intermediate goal which the manipulator moves to first to allow it to move around the obstacle in another direction.

7.3.2 Cell Decomposition Approach

In the cell decomposition approach, the end-effector essentially moves from cell to cell in the same fashion as the particle in Section (7.2.2). But, a cell is also considered to be unsuitable if no suitable joint variables can be found (even if they do exist - the GA may not find them). This approach overcomes the above problem with the manipulator becoming trapped by an obstacle blocking the proximal links. The grid is aligned to the start, and the goal may not be aligned. If the goal is not aligned on the grid then the end-effector moves until it is less than δd from it and then the next target is the goal.

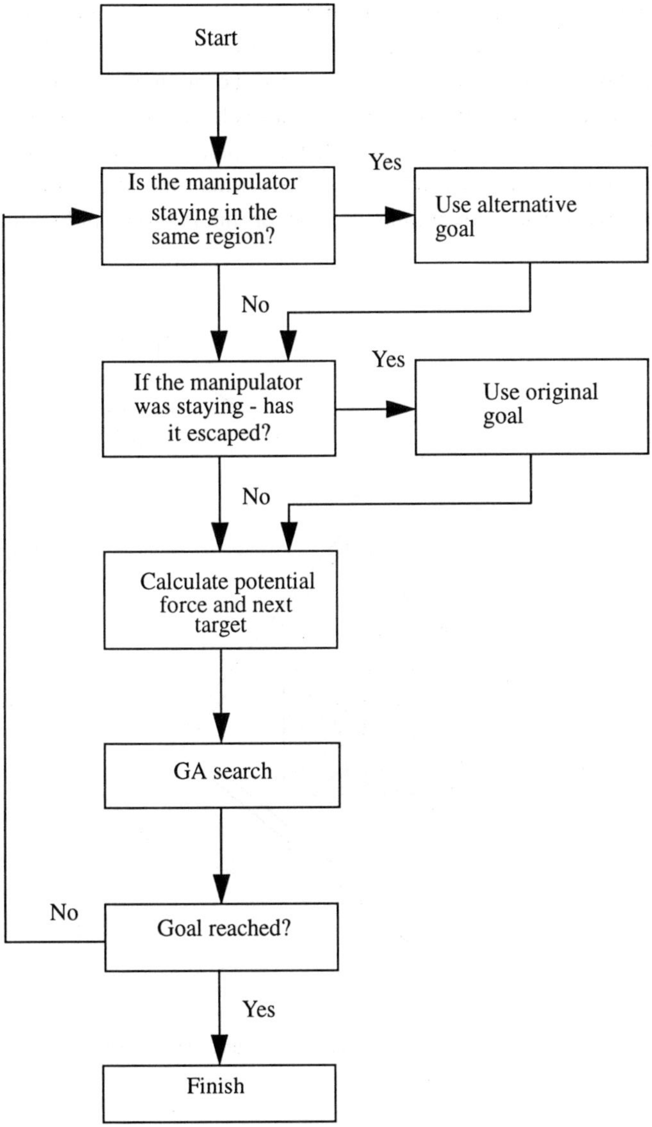

Figure 7.14: Outline of the potential field algorithm

The problem with the cell decomposition approach is that a moving obstacle may block the path of a manipulator in one direction but later the manipulator will be free to move in that direction, but it will not know this. Once a cell is marked as occupied it stays marked. The manipulator may be in another path of the workspace and will not be able to detect that the obstacle has moved. A possible solution to this may be to retry some paths if the manipulator has been unsuccessful in reaching the goal and if it was aware that there where moving obstacles in some regions of the workspace. An outline of the algorithm is shown in Figure (7.16).

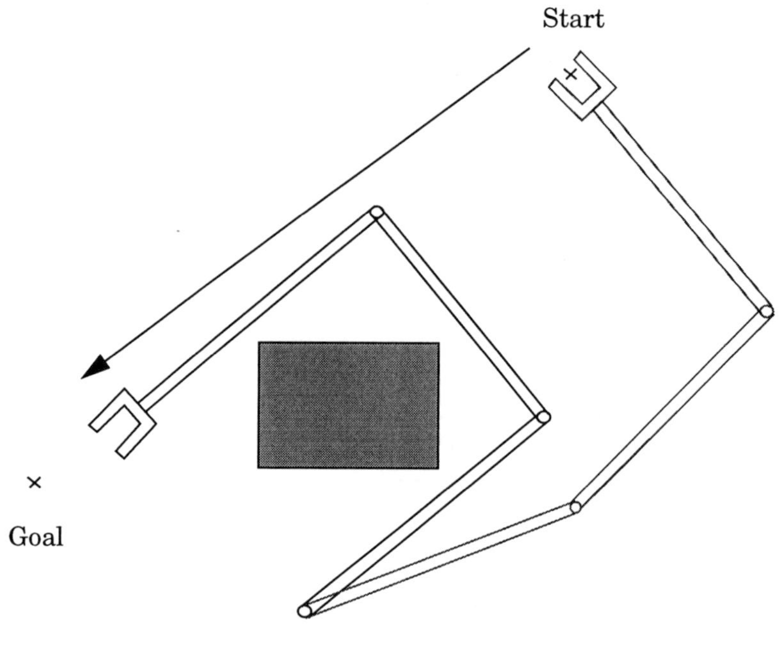

Figure 7.15: Case where manipulator is unable to proceed

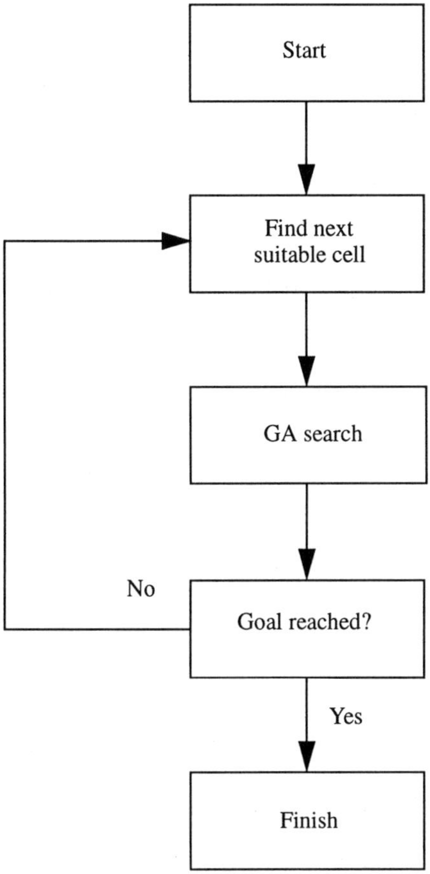

Figure 7.16: Outline of the cell decomposition algorithm

7.3.3 Genetic Algorithm Details

The GA has an extra term added to the fitness function - an error term depending on the proximity of the manipulator to nearby obstacles. The fitness is evaluated in two ways: How close the end-effector is to the desired position? And, how far away from other obstacles the manipulator is? If the manipulator collides with an obstacle for this particular set of joint variables then the fitness of that particular string is set to zero.

The fitness function (similar to Equation 4.11) is defined as:

$$f = \frac{1}{1 + w_1 \times e_1 + w_2 \times e_2} \tag{7.9}$$

The first error term is the positioning error, which is defined earlier in Equation (4.12).

The second error term for potential fields is the proximity error:

$$e_2 = \underset{i \in \{\text{obstacles}\}}{\text{Mean}} \left(\left\| \mathbf{F}_i \right\| \right) \tag{7.10}$$

The second error term for cell decomposition is the proximity error:

$$e_2 = \underset{i \in \{\text{obstacles}\}}{\text{Mean}} \left(\frac{1}{D_i} \right) \tag{7.11}$$

The normalized distance from the obstacles in its proximity is:

$$D_i = \begin{cases} \left(\vartheta - d_i \right) / R, & d_i \leq \vartheta \\ 0, & d_i > \vartheta \end{cases} \tag{7.12}$$

Where: w_2 = Second error weight, and
$\mathbf{F}_i$ = Repulsive force from obstacle i to the closest point on the manipulator (7.7).
d_i = Minimum distance from manipulator to obstacle i,
R = Manipulator reach (4.13), and
J = Proximity threshold.

The above algorithm requires the calculation of the distances between obstacles and the manipulator and to know which point on the obstacle is closest. Various techniques exist, as mentioned in Chapter (5), but the method developed in Chapter (5) is used.

7.4 Conclusions

This chapter has presented an algorithm to solve the collision avoidance problem for robot manipulators. Several examples of its operation will be presented in the next chapter (Chapter 8). The approach used has built upon the GA solution to the inverse-kinematics problem in Chapter (5) with an additional penalty for the collision avoidance algorithm. The guidance of the end-effector is provided by potential fields and approximate cell decomposition. They form the main building block for the collision avoidance algorithm. Two trivial examples of them used to guide a particle successfully demonstrated their operation.

CHAPTER 8

EXAMPLES

8.1 Introduction

There are two platforms used. The first is an IBM compatible PC, on which the GA runs sequentially. The second is a network of transputers using the PC as a host and they run the parallel version of the GA.

There are seven examples presented, in each case the manipulators move successfully without collision from their starting points to their goals. The first three examples use potential fields and a serial GA. The next three examples use cell decomposition and a serial GA. The final example uses potential fields with a parallel GA. The parallel GA is used to speed up computations. In all the examples only the positioning of the end-effector is considered and not the orientation. All the manipulators have the DOF required to position the end-effector in the workspace plus an extra DOF for collision avoidance. In all the figures the view is down the Z axis.

8.2 Serial Examples

Tables (8.1), (8.2) and (8.3) contains the link parameters, Table (8.4) contains the GA and potential field parameters used, and Table (8.5) contains the summary of results for Examples (1), (2) and (3). Tables (8.6), (8.7) and (8.8) contains the link parameters, Table (8.9) contains the GA and cell decomposition parameters used, and Table (8.10) contains the summary of results for Examples (4), (5) and (6). In the link parameter tables, the symbol **V** means the joint is variable and the symbol **R** stands for a revolute joint as opposed to a prismatic joint.

Link	Link type	a_i (m)	α_i (rad)	d_i (m)	θ_i (rad)	Link thickness (m)	$q_{\min}$	$q_{\max}$
1	R	2	0	0	V	0.01	0	2π
2	R	2	0	0	V	0.01	0	2π
3	R	2	0	0	V	0.01	0	2π

Table 8.1: Link parameters for Example 1

Robot	Link	Link type	a_i (m)	α_i (rad)	d_i (m)	θ_i (rad)	Link thickness (m)	q_{min}	q_{max}
1	1	R	2	0	0	V	0.1	0	2π
1	2	R	2	0	0	V	0.1	0	2π
1	3	R	2	0	0	V	0.1	0	2π
2	1	R	2	0	0	V	0.1	0	2π
2	2	R	2	0	0	V	0.1	0	2π
2	3	R	2	0	0	V	0.1	0	2π

Table 8.2: Link parameters for Example 2

Robot	Link	Link type	a_i (m)	α_i (rad)	d_i (m)	θ_i (rad)	Link thickness (m)	q_{min}	q_{max}
1	1	R	0	$\pi/2$	0	V	0.1	0	2π
1	2	R	2	0	0	V	0.1	0	2π
1	3	R	1	0	0	V	0.1	0	2π
1	4	R	1	0	0	V	0.1	0	2π
2	1	R	0	$\pi/2$	0	V	0.1	0	2π
2	2	R	2	0	0	V	0.1	0	2π
2	3	R	1	0	0	V	0.1	0	2π
2	4	R	1	0	0	V	0.1	0	2π

Table 8.3: Link parameters for Example 3

Example	1	2	3
$\delta\omega$	0.1	0.1	0.1
ε	1	1	1
η	1	1	1
ρ_0	0.2	0.2	0.2
Population size	40	40	40
String length	24	36	40
p_c	0.7	0.7	0.7
p_m	0.01	0.01	0.01
Target error	1%	1%	1%
w_1	0.8	0.8	0.8
w_2	0.2	0.2	0.2
Loitering distance	0.05	0.05	0.05
Escape distance	0.3	1.0	0.4
Vicinity distance	0.4	0.5	0.5

Table 8.4: GA and potential parameters for Examples 1, 2, and 3

Example	1	2	3
Average error	0.6739%	0.5410%	0.7785%
Average iterations	11.12	9.24	36.97

Table 8.5: Summary of results for Examples 1, 2 and 3

8.2.1 Example 1

In the first case (two dimensions), see Figure (8.1), a planar manipulator with three revolute joints moves in between two obstacles. Every tenth step is shown on the figure. This example demonstrates the ability of the algorithm to manoeuvre a manipulator in a narrow space.

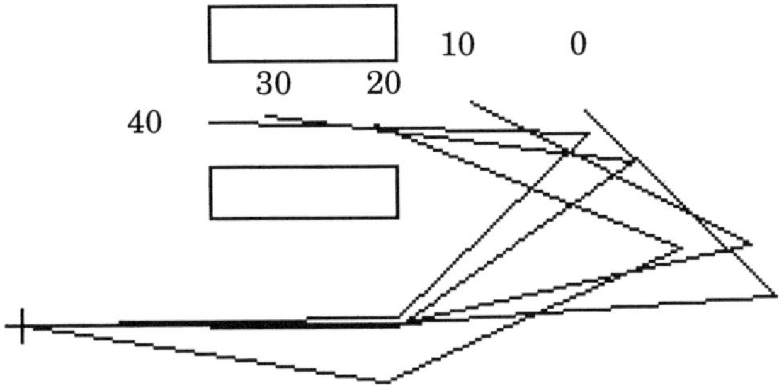

Figure 8.1: Example 1

8.2.2 Example 2

In the second case (two dimensions), see Figure (8.2), two planar manipulators, each with three revolute joints, move around each other to their goals. Every fourteenth step is shown in the figure. The two manipulators move towards each other and then are diverted towards their base until they are out of the influence of each other and then continue on their way to their respective goals.

8.2.3 Example 3

In the third case (three dimensions), see Figure (8.3), two manipulators, each with four revolute joints move to their goals while avoiding a mobile object (moving from O_S to O_F). Here, each manipulator M_i ($i = 1$ or 2) moves from S_i to F_i along path P_i. The view shown in Figure (8.3) is slightly deceptive - the first manipulator moves over the top of the object. The first manipulator is diverted towards its base when it approaches the obstacle. This is apparent in the end-effector path traced out on the figure.

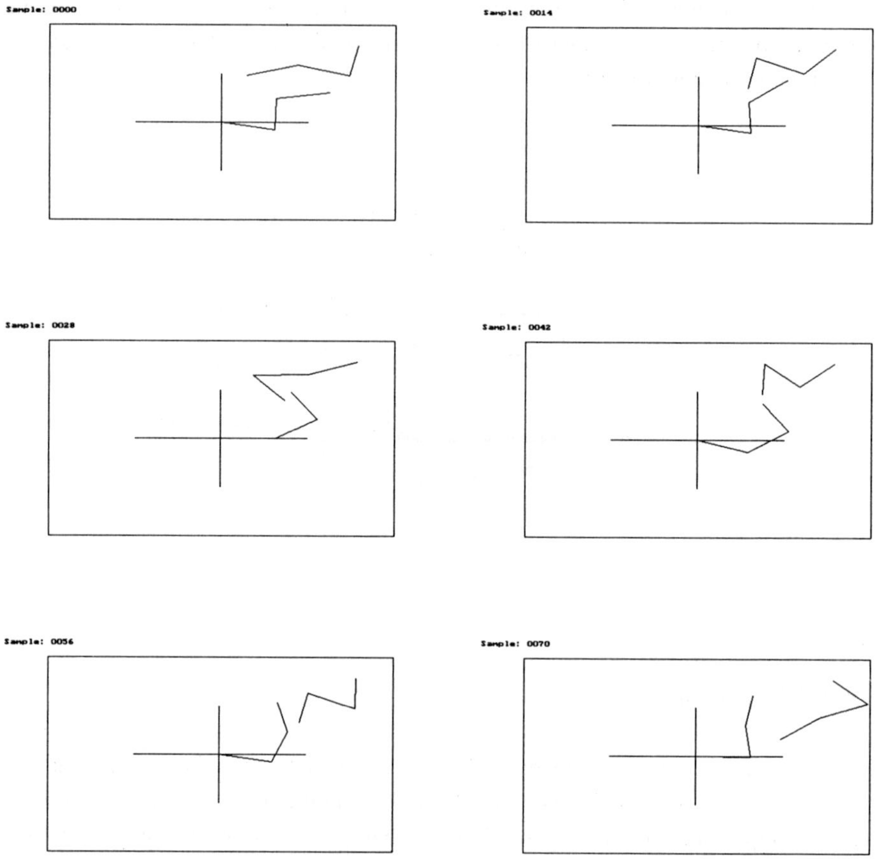

Figure 8.2: Example 2

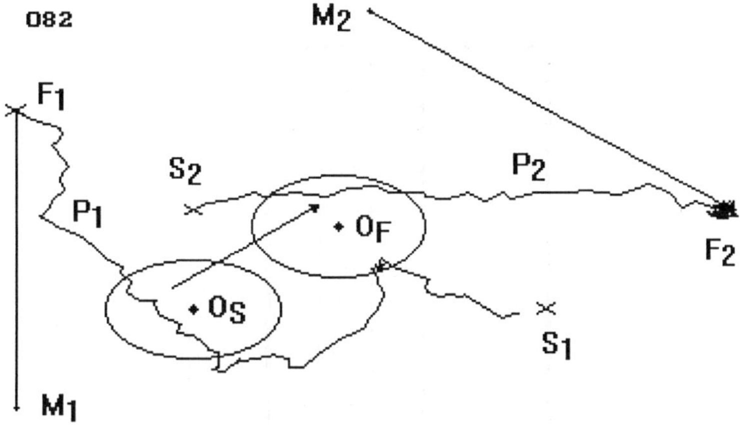

Figure 8.3: Example 3

Link	Link type	a_i (m)	α_i (rad)	d_i (m)	θ_i (rad)	Link thickness (m)	$q_{\min}$	$q_{\max}$
1	R	1	0	0	V	0.01	0	2π
2	R	1	0	0	V	0.01	0	2π
3	R	1	0	0	V	0.01	0	2π

Table 8.6: Link parameters for Example 4

Link	Link type	a_i (m)	α_i (rad)	d_i (m)	θ_i (rad)	Link thickness (m)	q_{min}	q_{max}
1	R	1	0	0	V	0.01	0	2π
2	R	1	0	0	V	0.01	0	2π
3	R	1	0	0	V	0.01	0	2π

Table 8.7: Link parameters for Example 5

Robot	Link	Link type	a_i (m)	α_i (rad)	d_i (m)	θ_i (rad)	Link thickness (m)	$q_{\min}$	$q_{\max}$
1	1	R	0	$\pi/2$	0	V	0.1	0	2π
1	2	R	2	0	0	V	0.1	0	2π
1	3	R	1	0	0	V	0.1	0	2π
1	4	R	1	0	0	V	0.1	0	2π
2	1	R	0	$\pi/2$	0	V	0.1	0	2π
2	2	R	2	0	0	V	0.1	0	2π
2	3	R	1	0	0	V	0.1	0	2π
2	4	R	1	0	0	V	0.1	0	2π

Table 8.8: Link parameters for Example 6

Example	4	5	6
Grid size	4x4	4x4	7x7x5
Cell size	0.1	0.1	0.2
Population	50	50	60
String length	30	30	60
p_c	0.7	0.7	0.7
p_m	0.01	0.01	0.01
Target error	1%	1%	2%
w_1	0.8	0.8	0.8
w_2	0.2	0.2	0.2
ϑ	0.1	0.1	0.1

Table 8.9: GA and cell parameters for Examples 4, 5 and 6

Example	4	5	6
Average error	0.662%	0.772%	2.17%
Average number of iterations	38.00	56.73	116.88

Table 8.10: Summary of results for Examples 4, 5, and 6

8.2.4 Example 4

In the fourth case a three DOF planar manipulator moves from its starting point to the goal without collision, refer to Figure (8.4). Every fourth step is shown. This example demonstrates the backtracking capability of the algorithm. When the manipulator reaches a dead-end formed by the obstacles it backtracks to move around the obstacle in the other direction.

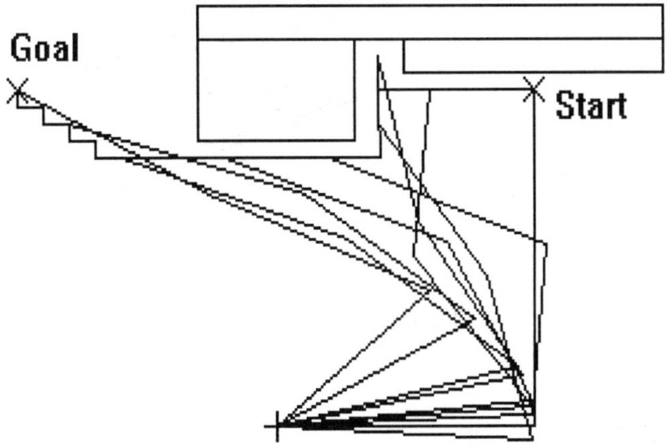

Figure 8.4: Example 4

8.2.5 Example 5

In the fifth case a three DOF planar manipulator moves from its starting point to the goal without collision, refer to Figure (8.5). Every sixth step is shown on the figure. The example highlights the cell-to-cell motion of the end-effector. The manipulator moves in a staircase manner around the obstacle. This apparently erratic motion is due to the cell decomposition guidance, which moves the end-effector to one of four adjacent cells in two dimensions (there would be a choice of six cells in three dimensions).

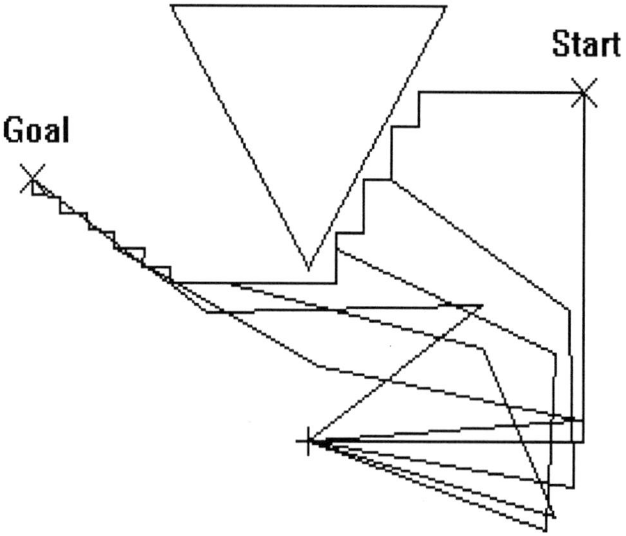

Figure 8.5: Example 5

8.2.6 Example 6

In the sixth case (Figure 8.6) two manipulators, each with four DOF, move around each other in three dimensions to avoid collision. Every eighth step is shown in the figure. Even though the manipulators managed to reach their goals without collision, the average error was greater than the target error. The goals were chosen to place the manipulators on a collision course - which the algorithm was able to overcome, enabling the manipulators to move to their goals without collision.

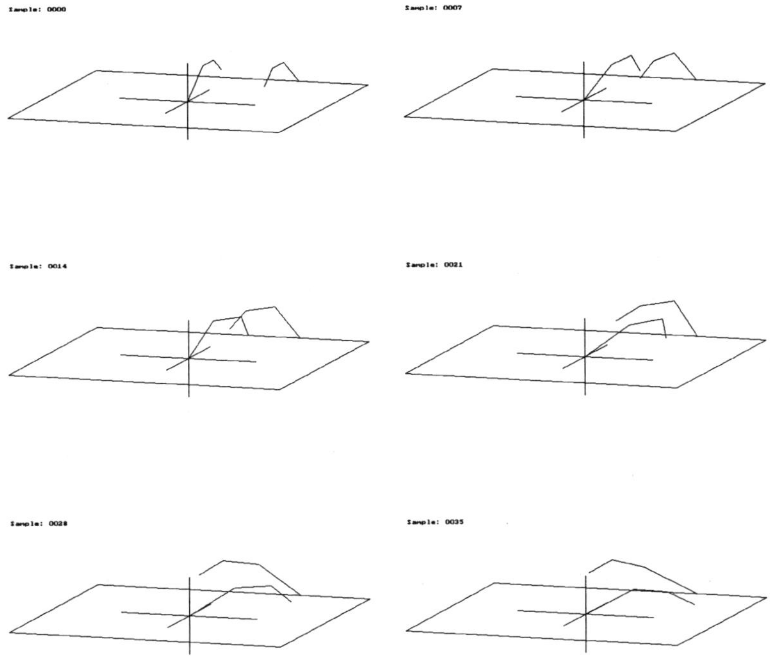

Figure 8.6: Example 6

8.3 Parallel Example

Table (8.11) contains the link parameters, Table (8.12) contains the GA and potential field parameters used, and Table (8.13) contains the summary of results for Example (7).

The seventh case is similar to the first, except that it is implemented on transputers to examine the performance of parallel processing, The number of transputers used ranged from one to six. If the number of iterations per processor is considered, the cases with more iterations per processor took longer as expected. The correlation coefficient between the time and iterations per processor is approximately 92.7%. But as the number of processors increased the time taken to perform an iteration dropped as expected showing that increasing the number of processors does directly increase the performance, the correlation coefficient for this is -84.2%. This is reflected in the speed-up, *S*, from (2.1) in Table (8.13). But the time taken per iteration per processor shows a drop in efficiency as the number of processors increases (even though it does take less time), this is due to the communication overheads. This is reflected in the efficiency, f, from (2.5) in Table (8.13).The correlation coefficient for this drop in efficiency is approximately 99.0%. These results are displayed in graphs in Figures (8.7) - (8.9).

Link	Link type	a_i (m)	α_i (rad)	d_i (m)	θ_i (rad)	Link thickness (m)	$q_{\min}$	$q_{\max}$
1	R	2	0	0	V	0.01	0	2π
2	R	2	0	0	V	0.01	0	2π
3	R	2	0	0	V	0.01	0	2π

Table 8.11: Link parameters for Example 7

$\delta\omega$	0.1
ε	1
η	1
ρ_0	0.2
Population	40
String length	21
p_c	0.7
p_m	0.01
Target error	1%
w_1	0.8
w_2	0.2
Loitering distance	0.05
Escape distance	0.3
Vicinity distance	0.4

Table 8.12: GA and potential parameters for Example 7

Slaves	Time (sec)	Total number of iteration s	Average error	Time per iteration	Iterations per processor	Time per iteration per processor	S	ϕ
1	207.29	167	0.66%	1.24	167	1.241	1.000	1.000
2	193.72	278	0.72%	0.701	139	1.393	1.768	0.884
3	124.39	204	0.68%	0.609	68	1.829	2.036	0.679
4	115.34	203	0.57%	0.568	50.75	2.273	2.185	0.546
5	171.31	356	0.65%	0.481	71.2	2.406	2.580	0.516
6	99.31	209	0.69%	0.475	34.83	2.851	2.612	0.435

Table 8.13: Summary results for Example 7

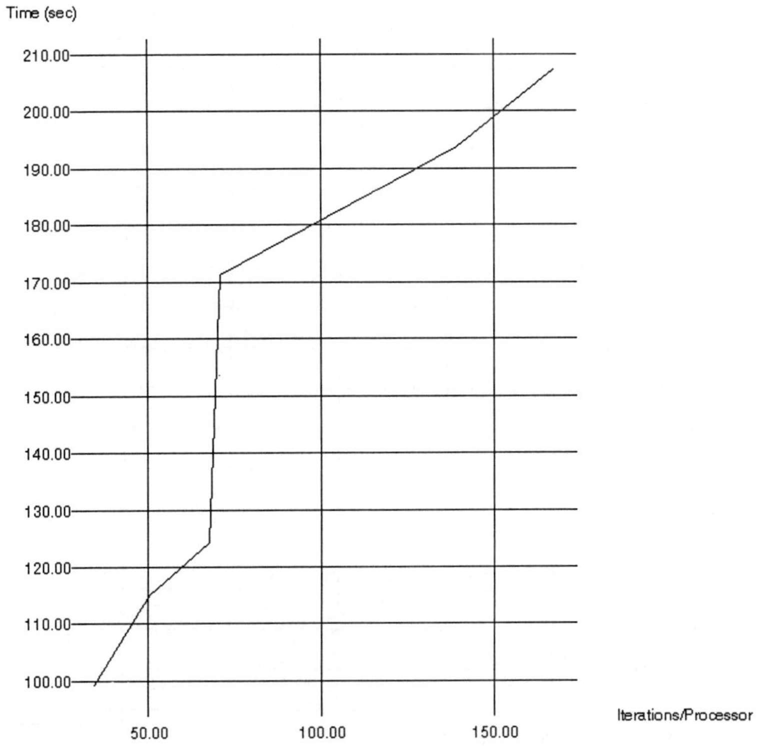

Figure 8.7: Time vs Iterations per processor

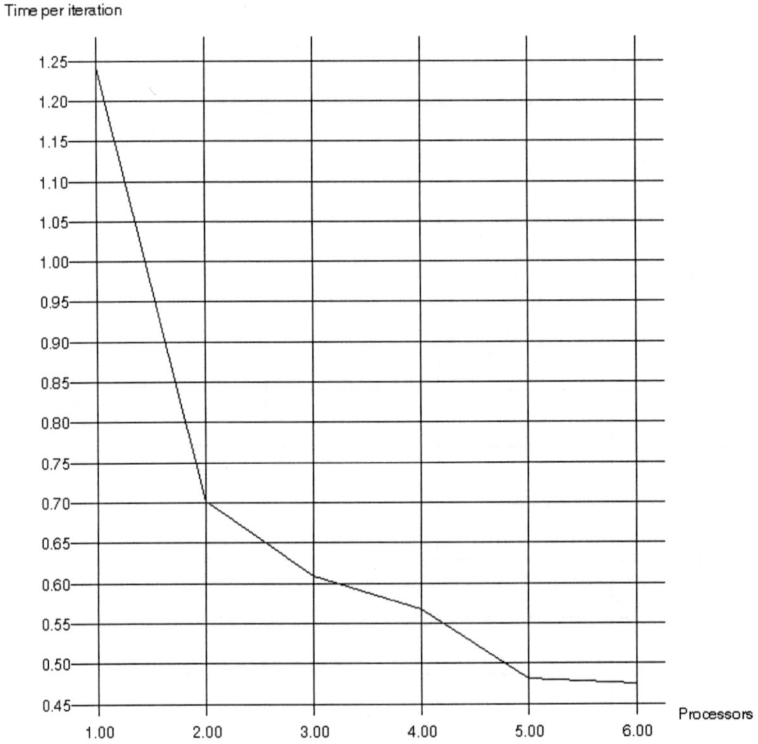

Figure 8.8: Processors vs Time per iteration

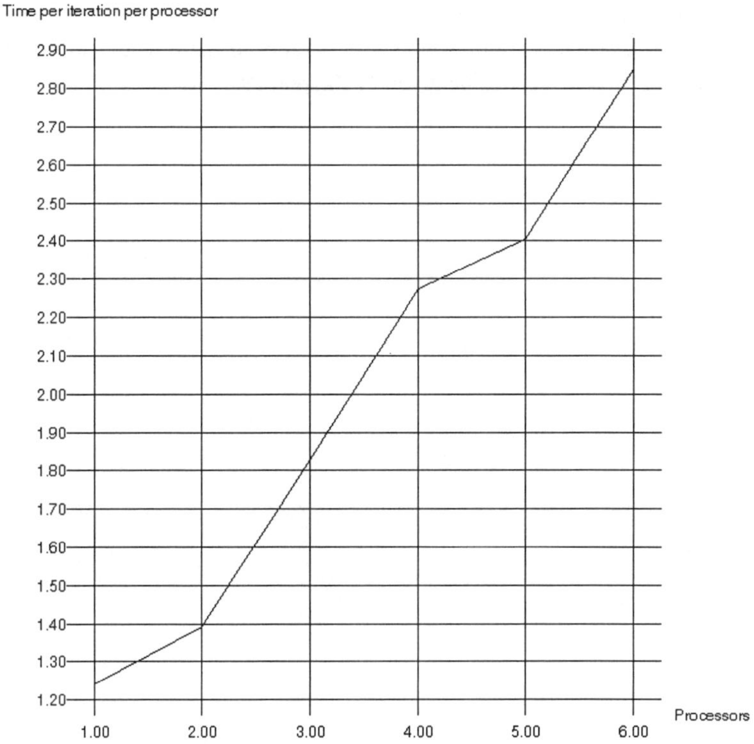

Figure 8.9: Processors vs Time per iteration per processor

8.4 Conclusions

The algorithm demonstrated in this chapter successfully generated collision free trajectories to enable multiple robot manipulators to achieve their goals in an environment with obstacles (stationary and mobile). The manipulators only possessed limited knowledge of their environment (the distances to the closest objects and the location of the closest points). The method does not use a configuration space mapping but instead operates in cartesian space. The performance of the algorithm was greatly improved by the application of a parallel GA. The GA used has taken effective advantage of parallel processing to greatly speed up its operation. Each fitness calculation for a string is computationally expensive, and the whole population of strings could be evaluated simultaneously in parallel, if there where sufficient processors. Further discussion will be given in Chapter (9).

CHAPTER 9

DISCUSSION, CONCLUSIONS AND FUTURE WORK

9.1 Introduction

The path planning problem is one important aspect of robot control. The primary objective of this work was to develop a collision avoidance algorithm for multiple robot manipulators in the same environment with other obstacles. The other obstacles can be stationary, mobile or shape-changing objects. The method is a general technique that can be applied to any manipulator. The method developed can be used as an on-line approach since it does not require a complete description of the workspace and uses local information as the manipulators progress towards their goals.

Robots are systems with hard real-time constraints to be met. They have an enormous amount of calculations to be performed, and these calculations must be performed in a very short time interval, due to the high sampling rates used. Parallel processing has been used to speed up the performance of the algorithm. The GA itself is parallel in nature and has been implemented on a parallel architecture to improve its operating speed.

9.2 Summary of Results and Discussion

A GA solution to the inverse kinematics problem was developed. The GA is able to search for a suitable solution to position an end-effector in any realizable position. Each genetic string represented a particular joint configuration and then the forward kinematics equation is then evaluated to test where the end-effector would be positioned. The fitness of the string was dependent on the positioning error. The examples used demonstrated its operation. The GA used has taken effective advantage of parallel processing to greatly speed up its performance, each fitness calculation for a string is computationally expensive and the whole population of strings could be evaluated simultaneously in parallel if there where sufficient processors.

The collision detection problem was solved by measuring the distances between a manipulator and other obstacles in the environment. A model of the manipulator was developed based on the link parameters. The model is arbitrary and can be used for any manipulator following the DH convention. The objects in the workspace are represented by spheres and plane segments. The simplified model results in less calculations and therefore faster processing. The algorithm used can take effective advantage of parallel processing to speed up the calculations, especially where there are a large number of manipulator links and/or spheres or plane segments. This is due to the amount of independent calculations that need to be performed (i.e. computing the distance from each link of the manipulator to each obstacle in the environment). The examples used again demonstrate its successful operation.
In the final part of the work a collision avoidance algorithm was developed and implemented by computer simulation. The algorithm was successful in generating collision-free trajectories for multiple robot manipulators exploring their environment with obstacles. The algorithm used a modified GA solution to the inverse kinematics problem, along with the collision detection method and either potential fields or approximate cell decomposition. Each manipulator explores the region of space in its vicinity and attempts to move closer to the goal (without colliding with any obstacles) during each step.

The fitness function in the GA solution to the inverse kinematics problem had an additional error term added to it. The collision detection algorithm was used to calculate the distances from manipulators to obstacles and return the closest points on the manipulator and the obstacle. The potential fields and approximate cell decomposition were used to provide the guidance for the end-effector as it moved through cartesian space.

The specific advantages of the approach used are:

- The method does not require a mapping into configuration space saving a lot of processing time. The configuration space approach also requires complete knowledge of the workspace and if it is time-varying (i.e. mobile or shape changing obstacles are present) then it will need to be constantly recomputed. Most of the current approaches use configuration space. The complexity (in terms of dimension) of the configuration space is proportional to the DOF of the manipulator, but in cartesian space the dimension is fixed.

- The algorithm uses a GA which can find solutions to problems rapidly, but is not guaranteed to find a solution. The completeness is traded off for greater computational speed. In the cases presented here it found satisfactory solutions.

- The algorithm is not limited by the DOF of the manipulators, the number of manipulators or objects (or even type of objects in the workspace) in the workspace (i.e. it is robust). But, in a very cluttered environment the algorithm may fail. The potential field and cell decomposition approaches have their respective drawbacks.

- Parallel processing can be effectively used to improve the performance of the system. This is apparent in the parallel example.

- The algorithm can be used in on-line path planning for a manipulator in real-time in an environment with unknown obstacles but the following points will need to be addressed: (1) Being able to detect when it is near obstacles (similar to the distance of influence in potential fields). It does not require a complete description of all the obstacles (or any prior knowledge of any of them), as a configuration space approach would, only local knowledge is required. A proximity sensor (e.g. sonar) may address this issue. (2) Very fast processing to enable the manipulator to move with reasonable speed. Taking effective advantage of parallel processing can solve this. And, (3) the ability to examine the feasibility of the next position before moving there.

9.3 Limitations

Even though the approach used has certain advantages, it also has limitations. These limitations can form the basis for further developments and future work. The main limitations are:

- In each of the collision avoidance examples a solution was found, but the solutions may not have been optimal (shortest path). To locate the shortest path an algorithm would require complete knowledge of the workspace and then locate the most optimal path off-line using a search strategy. If the manipulator was to repeat a particular motion and then if it remembered its old path it may be able to improve upon it. This may only work in a static environment where everything, apart from the manipulator, remains unchanged.

- The primary drawback of not using a configuration space approach is that the inverse kinematics problem must be solved, which, as mentioned earlier, can be difficult. The level of difficulty is proportional to the DOF of the manipulator. But the GA manages to find satisfactory solutions.

- The approach used only finds a suitable path without considering the dynamic properties of the manipulator or the effect that the dynamics may have on the motion. Path planning is just one component of robot control and it is what this work focussed on.

- The potential field guidance suffered from local minima which can trap the manipulator. This was overcome by temporarily diverting it in another direction to enable the manipulator to move until it was out of the influence of whatever obstacle caused the local minima. The manipulator could still become trapped by its proximal links, as shown in Chapter (7). The approximate cell decomposition approach did not allow the manipulator to become trapped like potential fields did, but it could waste a lot of time searching if it took a wrong turn and then had to eventually backtrack. The approximate cell decomposition approach will attempt to consider all possibilities. If there were mobile obstacles in the workspace it is possible that they could cut off the return route during backtracking.

- The selection of the potential field or cell decomposition parameters and the GA parameters will affect the performance of the algorithm. The selection of these parameters is often a process of trial and error.

- The main drawback in both the potential field and cell decomposition approaches was that the path followed was not smooth. The potential field always produced a smooth path while the direction of the force was constant, but it was very jagged when the end-effector bounced in and out of the distance of influence of an obstacle. The path located in the cell decomposition approach could be smoothed to some degree by allowing moves to diagonal neighbors rather than the ones directly on the side.

9.4 Future Work

The limitations mentioned in the previous section can form the basis for future work. Any extensions of this work can be in the areas of practical application and further theoretical investigation. In particular the following points could be considered:

- Implementing the algorithm in a real robotic environment, while considering the issues mentioned in Section (9.2).
- Using other techniques to solve the inverse kinematics problem, such as simulated annealing and comparing the performance with GAs.
- To make the collision avoidance algorithm more practical it should be incorporated into a complete task planning system to produce the input for the robot controller.
- Further work in the implementation onto parallel processing architectures can be done to improve its performance even further.
- Improving the algorithm to overcome some of the problems associated with the potential field and cell decomposition guidance used. Other ways of guiding the end-effector could even be considered and compared.

9.5 Conclusions

The algorithm used here worked well in all the cases examined, but it could possibly be improved to overcome some of the problems encountered. The accuracy depends on how close the robot model is to the real robot manipulator, how the radius of each link is chosen (an average value or the maximum possible value), and how close the object model is to the real object. How well they perform depends on how the GA parameters and the potential field or approximate cell decomposition parameters are chosen. The sampling rate also needs to be high enough so that no collisions are missed. This requires very fast processing. The processing speed has been increased by using parallel processing.

To properly asses the performance of the path planner it must be compared with other works in path planning. A review of some of the previously developed path planners is given in Chapter (3), along with their key features. Most of them are limited in their application, operate off-line and use configuration space. The approach developed here is general (in that it can be applied to a wide range of problem types) and is suited to on-line implementation. The specific advantages have been discussed earlier.

One point to note that the real speed of the algorithm is dependent on the computer system used. Some of the earlier algorithms may have been limited by slower computers so the computational speed may be slower than what it could be on the faster computers currently used today.

The ultimate test of the algorithms effectiveness would be to implement it as part of the task planner in a real robotic system.

REFERENCES

AHUACTZIN, J. M., TALBI, E. -G., BESSIERE, P. and MAZER, E. (1992). "Using Genetic Algorithms for Robot Motion Planning," in *Proceedings of the Tenth European Conference on Artificial Intelligence*, Vienna, August 3 - 7, pp. 671 - 675.

AKL, S. G. (1989). *The Design and Analysis of Parallel Algorithms*, Prentice-Hall, Englewood Cliffs, New Jersey.

ALANDER, J. T. (1991). "On Finding the Optimal Genetic Algorithms for Robot Control Problems," in *Proceedings of the IEEE/RSJ International Workshop on Intelligent Robots and Systems*, Osaka, November 3 - 5, pp. 1313 - 1318.

ANKENBRANDT, C. A., BUCKLES, B. P. and PETRY, F. E. (1990). "Scene Recognition using Genetic Algorithms with Semantic Nets," *Pattern Recognition Letters*, Vol. 11, No. 4 (April), pp. 285 - 293.

BARON, R. J. and HIGBIE, L. (1992). *Computer Architecture Case Studies*, Addison-Wesley, Reading, Massachusetts.

BARRAQUAND, J. and LATOMBE, J. -C. (1990). "A Monte-Carlo Algorithm for Path Planning with Many Degrees of Freedom," in *Proceedings of the IEEE International Conference on Robotics and Automation*, Cincinnati, May 13 - 18, pp. 1712 - 1717.

BARRAQUAND, J. and LATOMBE, J. -C. (1991). "Robot Motion Planning: A Distributed Representation Approach," *International Journal of Robotics Research*, Vol. 10, No. 6 (December), pp. 628 - 649.

BARRAQUAND, J., LANGLOIS, B. and LATOMBE, J. -C. (1992). "Numerical Potential Field Techniques for Robot Path Planning," *IEEE Transactions on Systems, Man, and Cybernetics*, Vol. 22, No. 2 (March/April), pp. 224 - 241.

BASTA, R. A., MEHROTRA, R. and VARANASI, M. R. (1988). "Detecting and Avoiding Collisions Between Two Robot Arms in a Common Workspace," in *Robot Control: Theory and Applications*, Warwick, K. and Pugh, A. ed., Peter Peregrinus Ltd., London, pp. 185 - 192.

BESSIERE, P., AHUACTZIN, J. -M., TALBI, E. G. and MAZER, E. (1993). "The Ariadne's Clew Algorithm: Global Planning with Local Methods," in *Proceedings of the IEEE Conference Intelligent Robots and Systems*, Yokohama.

BETHKE, A. (1981). "Genetic Algorithms as Function Optimizers," *Ph.D. thesis, University of Michigan.*

BOISSIERE, P. T. and HARRIGAN, R. W. (1988). "Telerobotic Operation of Conventional Robot Manipulators," in *Proceedings of the IEEE International Conference on Robotics and Automation*, Philadelphia, April 24 - 29, pp. 576 - 583.

BONNEY, M. C., DOONER, M., TAYLOR, N. K., GREEN, J. L. and HEIGINBOTHAM, W. B. (1983). "Verifying Robot Programs for Collision Free Tasks," in *Developments in Robotics,* Rooks, B. ed., IFS Publications, Amsterdam, North Holland, pp. 257 - 263.

BRADY, M. (1989). *Robotics Science*, MIT Press, Cambridge, Massachusetts.

BROOKS, R. A. (1983). "Solving the Findpath Problem by Good Representation of Free Space," *IEEE Transactions on Systems, Man, and Cybernetics*, Vol. 13, No. 3 (March/April), pp. 190 - 197.

BUCKLES, B. P., PETRY, F. E. and KUESTER, R. L. (1990). "Schema Survival Rates and Heuristic Search in Genetic Algorithms", in *Proceedings of the Second IEEE International Conference on Tools for Artificial Intelligence*, Herndon, November 6 - 9, pp. 332 - 327.

BUCKLEY, S. J. (1989). "Fast Motion Planning for Multiple Moving Robots," in *Proceedings of the IEEE International Conference on Robotics and Automation*, Scottsdale, May 14 - 19, pp. 322 - 326.

CAMERON, S. (1989). "Efficient Intersection Tests for Objects Defined Constructively," *International Journal of Robotics Research*, Vol. 8, No. 1 (February), pp. 3 - 25.

CAMERON, S. (1990). "Collision Detection by Four-Dimensional Intersection Testing," *IEEE Transactions on Robotics and Automation*, Vol. 6, No. 3 (June), pp. 291 - 302.

CANNY, J. F. and REIF, J. (1987). "New Lower Bound Techniques for Robot Motion Planning Problems," in *Proceedings of the 28th Annual Symposium on Foundations of Computer Science*, Los Angeles, October 12 - 14, pp. 49 - 60.

CARRIKER, W. F., KHOSLA, P. K. and KROGH, B. H. (1990). "The Use of Simulated Annealing to Solve the Mobile Manipulator Path Planning Problem," in *Proceedings of the IEEE International Conference on Robotics and Automation,* Cincinnati, May 13 - 18, pp. 204 - 209.

CELA, A., HAMAM, Y. and GEORGES, D. (1991). "Decomposition Method for the Constrained Path Planning of Articulated Systems," in *Proceedings of the Fifth International Conference on Advanced Robotics: Robots in Unstructured Environments*, Vol. 2, Pisa, June 19 - 22, pp. 994 - 999.

CHAN, K. K. and ZALZALA, A. M. S. (1993). "Genetic-Based Minimum-Time Trajectory Planning of Articulated Manipulators with Torque Constraints," in *IEE Colloquim on Genetic Algorithms for Control Systems Engineering,* May 28, London, pp. 4/1 - 4/3.

CHEN, P. C. and HWANG, Y. K. (1992). "SANDROS: A Motion Planner with Performance Proportional to Task Difficulty," in *Proceedings of the IEEE International Conference on Robotics and Automation*, Nice, May 12 - 14, pp. 2346 - 2353.

CHEN, Y. C. and VIDYASAGAR, M. (1988). "Optimal Trajectory Planning for Planner n-Link Revolute Manipulators in the Presence of Obstacles," in *Proceedings of the IEEE International Conference on Robotics and Automation*, Philadelphia, April 24 - 29, pp. 202 - 208.

CHENG, G. -X., IKEGAMI, M. and TANAKA, M. (1992). "A Resistive Mesh Analysis Method for Parallel Path Searching," in *Proceedings of the 34th Midwest Symposium on Circuits and Systems*, Vol. 2, Monterey, May 14 - 19, pp. 827 - 830.

CHIEN, Y. P., KOIVO, A. J. and LEE, B. H. (1988). "On-Line Generation of Collision Free Trajectories for Multiple Robots," in *Proceedings of the IEEE International Conference on Robotics and Automation*, Philadelphia, April 24 - 29, pp. 209 - 211.

CHIPPERFIELD, A. and FLEMMING, P. (1996). "Parallel Genetic Algorithms," in *Parallel and Distributed Computing Handbook*, Zomaya, A. Y. ed., McGraw-Hill, New York, pp. 1118 - 1143.

COOMBS, S. and DAVIS, L. (1987). "Genetic Algorithms and Communication Link Speed Design: Constraints and Operators," in *Genetic Algorithms and their Applications: Proceedings of the Second International Conference on Genetic Algorithms*, Grefenstette, J. J. ed., Cambridge, July 28 - 31, pp. 257 - 260.

CORMEN, T. H., LEISERSON, C. E. and RIVEST, R. L. (1990). *Introduction to Algorithms*, MIT Press, Cambridge, Massachusetts.

CRAIG, J. J. (1986). *Introduction to Robotics: Mechanics and Control (2nd edition)*, Addison-Wesley, Reading, Massachusetts.

CRITCHLOW, A. J. (1985). *Introduction to Robotics*, Macmillan Publishing Company, New York.

CULLEY, R. K. and KEMPF, K. G. (1986). "A Collision Detection Algorithm based on Velocity and Distance Bounds," in *Proceedings of the IEEE International Conference on Robotics and Automation*, San Francisco, April 7 - 10, pp. 1064 - 1069.

DANIEL, J. W. (1981). *Elementary Linear Algebra and its Applications*, Prentice-Hall, Englewood Cliffs, New Jersey.

DENAVIT, J. and HARTENBERG, R. S. (1955). "A Kinematic Notation for Lower-Pair Mechanisms Based on Matrices," *ASME Journal of Applied Mechanics*, Vol. 22, No. 2 (June), pp. 215 - 221.

DENNIS, J. B. (1980). "Data Flow Supercomputers," *IEEE Computer*, Vol. 13, No. 11, pp. 48 - 56.

DORIGO, M. (1991). "Using Transputers to Increase Speed and Flexibility of Genetics-Based Machine Learning Systems," *Proceedings or the Euromicro Workshop on Real Time Systems*, Paris-Orsay, June 12 - 14.

ELTIMSAHY, A. H. and YANG, W. S. (1988). "Near Minimum-Time Control of Robotic Manipulator with Obstacles in the Workspace," in *Proceedings of the IEEE International Conference on Robotics and Automation*, Philadelphia, April 24 - 29, pp. 358 - 363.

ERDMANN, M. and LOZANO-PEREZ, T. (1986). "On Multiple Moving Objects," in *Proceedings of the IEEE International Conference on Robotics and Automation*, San Francisco, April 7 - 10, pp. 1419 - 1424.

FAVERJON, B. and TOURNASSOUND, P. (1987). "A Local Approach for Path Planning of Manipulators with a High Number of Degrees of Freedom," in *Proceedings of the IEEE International Conference on Robotics and Automation*, Raleigh, March 31 - April 3, pp. 1152 - 1159.

FAVERJON, B. (1989). "Hierarchical Object Models for Efficient Anti-Collision Algorithm," in *Proceedings of the IEEE International Conference on Robotics and Automation*, Scottsdale, May 14 - 19, pp. 333 - 340.

FIJANY, A. and BEJCZY, A. ed. (1992). *Parallel Computation Systems for Robotics: Algorithms and Architectures*, World Scientific, Singapore.

FILHO, J. L. R., TRELEAVEN, P. C. and ALIPPI, C. (1994). "*Genetic-Algorithm Programming Environments*," *in IEEE Computer*, Vol. 27, No. 6 (June), pp. 28 - 43.

FLYNN, M. J. (1966). "Very High-Speed Computing Systems," in *Proceedings of the IEEE*, Vol. 54, No. 12, pp. 1901 - 1909.

FORTUNE, S., WILFONG, G. and YAP, C. (1986). "Coordinated Motion of Two Robot Arms," in *Proceedings of the IEEE International Conference on Robotics and Automation*, San Francisco, April 7 - 10, pp. 1215 - 1233.

FOURMAN, M. P. (1985). "Compaction of Symbolic Layout using Genetic Algorithms," in *Proceedings of an International Conference on Genetic Algorithms and their Applications*, pp. 141 - 153.

FREUND, E., and HOYER, H. (1988). "Real-Time Pathfinding in Multirobot Systems Including Obstacle Avoidance," *International Journal of Robotics Research*, Vol. 7, No. 1 (February), pp. 42 - 70.

FUJIMURA, K. and SAMET, H. (1988). "Path Planning Among Moving Obstacles using Spatial Indexing," in *Proceedings of the IEEE International Conference on Robotics and Automation*, Philadelphia, April 24 - 29, pp. 1662 - 1667.

FUJIMURA, K. and SAMET, H. (1989). "Time Minimal Paths among Moving Obstacles," in *Proceedings of the IEEE International Conference on Robotics and Automation*, Scottsdale, May 14 - 19, pp. 1110 - 1115.

GE, Q. J. and MCCARTHY, J. M. (1990). "An Algebraic Formulation of Configuration-Space Obstacles for Spacial Robots," in *IEEE International Conference on Robotics and Automation*, Vol. 3, Issue 5, Cincinnati, May 13 - 18, pp. 1542 - 1547.

GILBERT, E. G., JOHNSON, D. W. and KEERTHI, S. S. (1988). "A Fast Procedure for Computing the Distance Between Complex Objects in Three-Dimensional Space," *IEEE Journal of Robotics and Automation*, Vol. 4, No. 2 (April), pp. 193 - 203.

GILL, M. A. C. and ZOMAYA, A. Y. (1995a). "On the Collision Detection Problem for Robot Manipulators," *Cybernetics and Systems*, Vol. 26, No. 2 (March - April), pp. 165 - 188.

GILL, M. A. C. and ZOMAYA, A. Y. (1995b). "Genetic Algorithms for Robot Control," in *Proceedings of the IEEE International Conference on Evolutionary Computation*, Perth, Western Australia, November 29 - December 1, pp. 462 - 466.

GILL, M. A. C. and ZOMAYA, A. Y. (1996). "Robot Manipulator Collision Avoidance Scheme," in *Proceedings of the Fourth International Conference on Control, Automation, Robotics and Vision*, Singapore, December 3 - 6, pp. 2194 - 2198.

GILL, M. A. C. and ZOMAYA, A. Y. (1998a). "A Parallel Collision Avoidance Algorithm for Robot Manipulators," *IEEE Concurrency* (in press).

GILL, M. A. C. and ZOMAYA, A. Y. (1998b). "A Cell Decomposition Based Collision Avoidance Algorithm for Robot Manipulators," in *Cybernetics and Systems* (in press).

GOLDBERG, D. E. (1983). "Computer-Aided Gas Pipeline Operation using Genetic Algorithms and Learning Rules," *Ph.D. thesis, University of Michigan.*

GOLDBERG, D. E. (1989). *Genetic Algorithms in Search, Optimization, and Machine Learning*, Addison-Wesley, Reading, Massachusetts.

GRENFENSTETTE, J. J. (1989). "A System for Learning Control Strategies with Genetic Algorithms," in *Proceedings of the third International Conference on Genetic Algorithms*, San Mateo, June 4 - 7, pp. 183 - 190.

GREFENSTETTE, J. J. and BAKER, J. E. (1989). "How Genetic Algorithms Work: A Critical Look at Implicit Parallelism," in *Proceedings of the third International Conference on Genetic Algorithms*, Schaffer, J. D. ed., San Mateo, pp. 12 - 19.

GUPTA, K. K. (1990). "Fast Collision Avoidance for Manipulator Arms: A Sequential Search Strategy," *IEEE Transactions on Robotics and Automation*, Vol. 4, No. 5, pp. 522 - 532.

HENNESSY, J. L.and PATTERSON, D. A. (1994). *Computer Organisation and Design: The Hardware/Software Interface*, Morgan Kaufmann Publishers, San Mateo, California.

HERMAN, M. (1986). "Fast, Three-Dimensional, Collision-Free Motion Planning," in *Proceedings of the IEEE International Conference on Robotics and Automation*, San Francisco, April 7 - 10, pp. 1056 - 1063.

HOGAN, N. (1985). "Impedance Control: An Approach to Manipulation: Part III - Application," *ASME Journal on Dynamic Systems, Measurement, and Control*, Vol. 107, March, pp. 17 - 24.

HOLLAND, J. H. (1975). *Adaptation in Natural and Artificial Systems: An Introductory Analysis with Applications to Biology, Control, and Artificial Intelligence*, The University of Michigan Press.

HOPCROFT, J., JOSEPH, D. and WHITESIDE, S. (1985). "On the Movement of Robot Arms in 2-Dimensional Bounded Regions," *SIAM Journal on Computing*, Vol. 14, No. 2 (May), pp. 315 - 333.

HWANG, K. (1993). *Advanced Computer Architecture: Parallelism, Scalability, and Programmability*, McGraw-Hill, New York.

HWANG, Y. K. and AHUJA, N. (1992). "Gross Motion Planning - A Survey," in *ACM Computing Surveys*, Vol. 24, No. 3 (September), pp. 219 - 291.

INMOS (1992). *The Transputer Databook*, SGS-Thomson Microelectronics, Italy.

JACAK, W. (1990). "Robot Task and Movement Planning," in *Proceedings of the AI, Simulation and Planning in High Autonomy Systems*, Tucson, March 26 - 27, pp. 168 - 173.

JANABI-SHARIFI, F. and VINKE, D. (1993). "Integration of the Artificial Potential Field Approach with Simulated Annealing for Robot Path Planning," in *Proceedings of the IEEE International Conference on Intelligent Control*, Chicago, August, pp. 536 - 541.

KAFRISSEN, E. and STEPHANS, M. (1984). *Industrial Robots and Robotics*, Reston Publishing Company, Reston, Virginia.

KANT, K. and ZUCKER, S. (1988). "Planning Collision-Free Trajectories in Time-Varying Environments: A Two-Level Hierarchy," in *Proceedings of the IEEE International Conference on Robotics and Automation*, Philadelphia, April 24 - 29, pp. 1644 - 1649.

KENTARNAVAZ, N. and LI, S. (1988). "A Collision-Free Navigation Scheme in the Presence of Moving Obstacles," in *Proceedings of the IEEE International Conference on Computer Vision*, Los Angeles, pp. 808 - 813.

KHATIB, O. and MAMPEY, L. M. (1978). "Fonction Decision-Commande d'un Robot Manipulator," Rep. 2/7156, DERA/CERT, Toulouse.

KHATIB, O. (1985). "Real-Time Obstacle Avoidance for Manipulators and Mobile Objects," in *Proceedings of the IEEE International Conference on Robotics and Automation*, St. Louis, pp. 500 - 505.

KHATIB, O. (1986). "Real-Time Obstacle Avoidance for Manipulators and Mobile Robots," *International Journal of Robotics Research*, Vol. 5, No. 1, pp. 90 - 98.

KHOOGAR, A. R. and PARKER, J. K. (1991). "Obstacle Avoidance of Redundant Manipulators Using Genetic Algorithms," in *IEEE Proceedings of Southeastcon*, Vol. 1, Williamsburg, April 7 - 10, pp. 317 - 320.

KHOSLA, P. and VOLPE, R. (1988). "Superquadratic Artificial Potentials for Obstacle Avoidance and Approach," in *Proceedings of the IEEE International Conference on Robotics and Automation*, Philadelphia, April 24 - 29, pp. 1778 - 1784.

KIM, J. and KHOSLA, P. K. (1992). "Real-Time Obstacle Avoidance Using Harmonic Potential Functions," *IEEE Transactions on Robotics and Automation*, Vol. 8, No. 3, pp. 338 - 349.

KLAFTER, R. D., CHMIELEWSKI, T. A. and NEGIN, M. (1989). *Robotic Engineering: An Integrated Approach*, Prentice Hall, Englewood Cliffs, New Jersey.

KONDO, K. (1991). "Motion Planning with Six Degrees of Freedom by Multistrategic Bidirectional Heuristic Free-Space Enumeration," *IEEE Transactions on Robotics and Automation*, Vol. 7, No. 3, pp. 267 - 277.

KWOK, D. P. and SHENG, F. (1994). "Genetic Algorithm and Simulated Annealing for Optimal Robot Arm PID Control," in *Proceedings of the IEEE Conference on Evolutionary Computation*, Orlando, June 27 - 29, pp. 707 - 713.

LATOMBE, J. -C. (1991). *Robot Motion Planning*, Kluwer Academic Publishers, Norwell, Massachusetts.

LAWSON, H. W. (1992). *Parallel Processing in Industrial Real-Time Applications*, Prentice Hall, Englewood Cliffs, New Jersey.

LEE, S. and PARK, J. (1991). "Cellular Robotic Collision-Free Path Planning," in *Proceedings of the IEEE fifth International Conference on Advanced Robotics - Robots in Unstructured Environments*, Pisa, June 19 - 22, pp. 539 - 544.

LEWIS, T. G. and EL-REWINI, H. (1992). *Introduction to Parallel Computing*, Prentice-Hall, Englewood Cliffs, New Jersey.

LIU, Y. H., KURODA, S., NANIWA. T., NOBORIO, H. and ARIMOTO, S. (1989). "A Practical Algorithm for Planning Collision-Free Coordinated Motion of Multiple Robots," in *Proceedings of the IEEE International Conference on Robotics and Automation*, Scottsdale, May 14 - 19, pp. 1427 - 1432.

LOZANO-PEREZ, T. and WESLEY, M. A. (1979). "An Algorithm for Planning Collision-Free Paths Among Polyhedral Obstacles," in *Communications of the ACM*, Vol. 22, No. 10, pp. 560 - 570.

LOZANO-PEREZ, T. (1983). "Spacial Planning: A Configuration Space Approach," *IEEE Transactions on Computers*, Vol. 32, No. 2, pp. 108 - 120.

LOZANO-PEREZ, T. (1986). "Motion Planning for Simple Robot Manipulators," in *Proceedings of the Third International Symposium on Robotics Research*, Gouvieux, October 7 - 11, pp. 133 - 140.

LOZANO-PEREZ, T. (1987). "A Simple Motion-Planning Algorithm for General Robot Manipulators," *IEEE Journal of Robotics and Automation*, Vol. 3, No. 3 (June), pp. 224 - 238.

LOZANO-PEREZ, T. and O'DONNELL, P. A. (1991). "Parallel Robot Motion Planning," in *Proceedings of the IEEE International Conference on Robotics and Automation*, Vol. 2, Sacramento, April 9 - 11, pp. 1000 - 1007.

LUMELSKY, V. (1986). "Continuous Motion Planning in Unknown Environment for a 3D Cartesian Robot Arm," in *Proceedings of the IEEE International Conference on Robotics and Automation*, San Francisco, April 7 - 10, pp. 1050 - 1055.

LUMELSKY, V. J. (1987). "Effect of Kinematics on Motion Planning for Planar Robot Arms Moving Amidst Unknown Obstacles," *IEEE Journal of Robotics and Automation*, Vol. 3, No. 3 (June), p. 207 - 223.

LUMELSKY, V. and SUN, K. (1987). "Gross Motion Planning for a Simple 3D Articulated Robot Arm Moving Amidst Unknown Arbitrarily Shaped Obstacles," in *Proceedings of the IEEE International Conference on Robotics and Automation*, Rayleigh, March 31 - April 3, pp. 1929 - 1934.

MACIEJEWSKI, A. A. and KLEIN, C. C. (1985). "Obstacle Avoidance for Kinematically Redundant Manipulators in Dynamically Varying Environments," *International Journal of Robotics Research*, Vol. 4, No. 3, pp. 109 - 117.

MANSOUR, N. and FOX, G. C. (1992). "Parallel Physical Optimization Algorithms for Data Mapping," in *Proceedings of the Second Joint International Conference on Vector and Parallel Processing*, Springer-Verlag, September 1 - 4, pp. 91 - 96.

MATTSON, T. G. (1996). "Scientific Computation," in *Parallel and Distributed Computing Handbook*, Zomaya, A. Y. ed., McGraw-Hill, New York, pp. 1118 - 1143.

MAZER, E., AHUCATZIN, J. M., TALBI, E., BESSIERE, P. and CHATROUX, T. "Parallel Motion Planning with the Ariadne's Clew Algorithm."

MEHROTRA, R., BASTA, R. A. and VARANASI, M. R. (1989). "Collision Detection between the Wrists of Two Robot Arms in a Common Workspace," *Journal of Intelligent and Robotic Systems*, Vol. 2, No. 1, pp. 29 - 41.

MICHALEWICZ, Z., KRAWCZYK, J. B., KAZEMI, M. and JANILOW, C. Z. (1990). "Genetic Algorithms and Optimal Control Problems," in *Proceedings of the Second IEEE International Conference on Decision and Control*, Honolulu, December 5 - 7, pp. 1664 - 1666.

MIYAZAKI, F. and ARIMOTO, S. (1984). "Sensory Feedback Based on the Artificial Potential for Robots," in *Proceedings of the 9th Annual ACM Symposium on Computational Geometry*, Berkley, June 6 - 8, pp. 63 - 72.

MOLDOVAN, D. I. (1993). *Parallel Processing: From Applications to Systems*, Morgan Kaufmann Publishers, San Mateo, California.

MUHLENBEIN, H. (1991). "Asynchronous Search by the Parallel Genetic Algorithm," in *Proceedings of the third IEEE International Symposium on Parallel and Distributed Processing*, Dallas, December 2 - 5, pp. 526 - 533.

MYERS, J. K. (1985). "A Robotic Simulator with Collision Detection: RCODE," in *Proceedings of the First Annual Workshop on Robotics and Expert Systems*, Houston, June 27 - 28.

NEWMAN, W. S. and HOGAN, N. (1987). "High Speed Robot Control and Obstacle Avoidance Using Dynamic Potential Function," in *Proceedings of the IEEE International Conference on Robotics and Automation*, Scottsdale, pp. 1104 - 1109.

NEWMAN, W. S., and BRANICKY, M. S. (1991). "Real-Time Configuration Space Transformations for Obstacle Avoidance," *International Journal of Robotics Research*, Vol. 10, No. 6 (December), pp. 650 - 667.

O'DONNELL, P. A. and LOZANO-PEREZ, T. (1989). "Deadlock-Free and Collision-Free Coordination of Two Robot Manipulators," in *Proceedings of the IEEE International Conference on Robotics and Automation*, Scottsdale, May 14 - 19, pp. 484 - 489.

PADEN, B., MEES, A. and FISHER, M. (1989). "Path Planning Using a Jacobian-Based Freespace Generation Algorithm," in *Proceedings of the IEEE International Conference on Robotics and Automation*, Vol 3., May 14 - 19, pp. 1732 - 1737.

PARKER, J. K., KHOOGAR, A. R. and GOLDBERG, D. E. (1989). "Inverse Kinematics of Redundant Robots using Genetic Algorithms," in *Proceedings of the IEEE International Conference on Robotics and Automation*, Vol. 1, Scottsdale, May 14 - 19, pp. 271 - 276.

PAUL, R. P. (1981). *Robot Manipulators: Mathematics, Programming, and Control: the Computer Control of Robot Manipulators*, MIT Press, Massachusetts, Cambridge.

PAVLOV, V. V. and VORONIN, A. N. (1984). "The Method of Potential Functions for Coding Constraints of the External Space in an Intelligent Mobile Robot," in *Soviet Auto. Cont. 6*.

PRASSLER, E. and MILIOS, E. (1990). "Parallel Distributed Robot Navigation in the Presence of Obstacles," in *Proceedings of the IEEE Symposium on Parallel and Distributed Computing*, Dallas, December 13 - 19, pp. 475 - 478.

RAMAMRITHAN, K. STANKOVIC, J. A., and ZHAO, W. (1989). "Distributed Scheduling of Tasks with Deadlines and Resource Requirements," *IEEE Transactions on Computers*, Vol. 38, No. 8, pp. 1110 - 1123.

REHG, J. (1985). *Introduction to Robotics: A Systems Approach*, Prentice-Hall, Englewood Cliffs, New Jersey.

REIF, J. H. and SHARIR, M. (1985). "Motion Planning in the Presence of Moving Obstacles," in *Proceedings of the 26th Annual IEEE Symposium on Foundations of Computer Science*, Portland, October 21 - 23, pp. 144 - 154.

RIMON, E. and KODITSCHEK, D. E. (1992). "Exact Robot Navigation using Artificial Potential Functions," *IEEE Transactions on Robotics and Automation*, Vol. 8, No. 5 (October), pp. 501 - 518.

SCHILLING, R. J. (1990). *Fundamentals of Robotics: Analysis and Control*, Prentice-Hall, Englewood Cliffs, New Jersey.

SCHWARTZ, J. T. and SHAHRIR, M. (1983). "On the Piano Movers' Problem: III Coordinating the Motion of Several Independent Bodies amidst Polygonal Barriers," *International Journal of Robotics Research*, Vol. 2, No. 3, pp. 46 - 75.

SCHWEIKARD, A. (1991). "Polynomial Time Collision Detection for Manipulator Paths Specified by Joint Motions," *IEEE Transactions on Robotics and Automation*, Vol. 7, No. 6 (December), pp. 865 - 870.

SEDGEWICK, R. (1990). *Algorithms in C*, Addison-Wesley, Reading, Massachusetts.

SHAHINPOOR, M. (1987). *A Robot Engineering Textbook*, Harper and Row Publishers, New York.

SHARIR, M. and SIFRONY, S. (1988). "Coordinated Motion Planning for Two Independent Robots," in *Proceedings of the 4th Annual ACM Symposium on Computational Geometry*, Urbana, June 6 - 8, pp. 319 - 328.

SHIBATA, T., FUKUDA, T., KOSUGE, K. and ARAI, F. (1992). "Selfish and Coordinative Planning for Multiple Mobile Robots by Genetic Algorithm," in *Proceedings of the 31st IEEE Conference on Decision and Control*, Tucson, December 16 - 18, pp. 2686 - 2691.

SHILLER, Z. and DUBOWSKY, S. (1988). "Global Time Optimal Motions of Robotic Manipulators in the Presence of Obstacles," in *Proceedings of the IEEE International Conference on Robotics and Automation*, Philadelphia, April 24 - 29, pp. 370 - 375.

SHILLER, Z. and DUBOWSKY, S. (1991). "On Computing the Global Time-Optimal Motions of Robotic Manipulators in the Presence of Obstacles," *IEEE Transactions on Robotics and Automation*, Vol. 7, No. 6 (December), pp. 785 - 797.

SIEGEL, H. J. (1990). *Interconnection Networks for Large-Scale Parallel Processing: Theory and Case Studies (2nd edition)*, McGraw-Hill, New York.

SNYDER, W. E. (1985). *Industrial Robots: Computer Interfacing and Control*, Prentice-Hall, Englewood Cliffs, New Jersey.

SOLANO, J. and JONES, D. I. (1993). "Generation of Collision-Free Paths, a Genetic Approach," in *Proceedings of the IEE Colloquium on GAs for Control Engineering*, London, May 28, pp. 5/1 - 5/6.

SOLANO, J. and JONES, D. I. (1994). "Parameter Determination for a Genetic Algorithm Applied to Robot Control," in *Proceedings of the IEE International Conference on Control*, London, March 21 - 24, pp. 765 - 770.

STONE, H. S. (1987). *High-Performance Computer Architecture*, Addison-Wesley, Reading, Massachusetts.

TALBI, E. -G. and MUNTEAN, T. (1993). "Hill-Climbing, Simulated Annealing and Genetic Algorithms: A Comparative Study and Application to the Mapping Problem," in *Proceedings of the IEEE International Conference on System Sciences*, Wailea, January 5 - 8, pp. 565 - 573.

TANESE, R. (1987). "Parallel Genetic Algorithm for a Hypercube," in *Genetic Algorithms and their Applications: Proceedings of the second International Conference on Genetic Algorithms*, Grefenstette, J. J. ed., Cambridge, July 28 - 31, pp. 177 - 183.

VAN LAARHOVEN, P. J. M. and Aarts, E. H. L. (1987). *Simulated Annealing: Theory and Applications*, D. Reidel Publishing Company, Dordrecht, Holland.

WARREN, C. W. (1989). "Global Path Planning using Artificial Potential Fields," in *Proceedings of the IEEE International Conference on Robotics and Automation*, Scottsdale, May 14 - 19, pp. 316 - 321.

WARREN, C. W. (1990). "Multiple Robot Path Coordination using Artificial Potential Fields," in *Proceedings of the IEEE International Conference on Robotics and Automation*, Cincinnati, May 13 - 18, pp. 500 - 505.

WATKINS, C. and SHARP, L. (1992). *Programming in Three Dimensions 3D Graphics, Ray Tracing, and Animation*, M & T Publishing Inc., San Mateo, California.

YEUNG, D. Y. and BEKEY, G. A. (1987). "A Decentralized Approach to the Motion Planning Problem for Multiple Robots," in *Proceedings of the IEEE International Conference on Robotics and Automation*, Rayleigh, March 31 - April 3, pp. 1799 - 1923.

YORK, T. A., PATRICK, D., STINCHCOMBE, M. S., TYRELL, S. G. and GREEN, P. R. (1994). "Sparc-Gap: a Parallel Genetic Algorithm Processing Platform," in *Proceedings of the IEE Colloquim on High Performance Applications of Parallel Architectures,* London, February 1, pp. 5/1 - 5/4.

YOSHIKAWA, T. (1990). "*Foundations of Robotics: Analysis and Control,*" MIT Press, Massachusetts, Cambridge.

ZHU, D. and LATOMBE, J. -C. (1991). "New Heuristic Algorithms for Efficient Hierarchical Path Planning," *IEEE Transactions on Robotics and Automation*, Vol. 7, No. 1 (February), pp. 9- 20.

ZOMAYA, A. Y. (1992). "*Modelling and Simulation of Robot Manipulators: a Parallel Processing Approach,*" World Scientific, Singapore.

ZOMAYA, A. Y. ed. (1996). *Parallel and Distributed Computing Handbook*, McGraw-Hill, New York.

APPENDIX 1

PARALLEL LINE PROOF

Norm: $\|v\| = \sqrt{v \cdot v} = \sqrt{v_x^2 + v_y^2 + v_z^2}$ (A1.1)

Inner product: $v_1 \cdot v_2 = \|v_1\| \|v_2\| \cos\Theta = \beta$ (A1.2)

Angle Q is the angle between the two vectors v_1 and v_2, with coincident starting points, as shown in Figure (A1.1). Two lines, with direction vectors v_1 and v_2, are parallel if:

$$\cos\Theta = \pm 1, \quad \text{(A1.3)}$$

i.e. $\Theta = k\pi;\ k \in Z.$ (A1.4)

If $\cos\Theta = \pm 1$, then:

$$|v_1 \cdot v_2| = \|v_1\| \|v_2\| \quad \text{(A1.5)}$$

i.e. $\beta = \sqrt{\alpha}\sqrt{\gamma}$ (A1.6)

i.e. $\alpha\gamma - \beta^2 = 0$ (A1.7)

Therefore $\Delta = 0$ (A1.8)

Refer to Section (6.3.1) for definition of symbols.

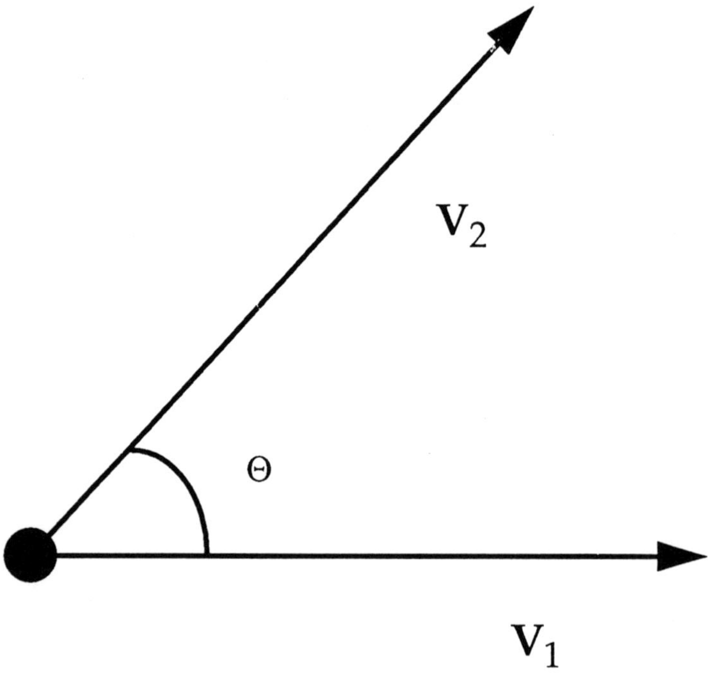

Figure A1.1: Angle between two vectors

APPENDIX 2

POLYNOMIAL METHOD

The polynomial method for joint trajectory calculation uses a fifth order polynomial (A2.1) to generate the trajectory of an individual joint. The joint displacement equation (A2.1) is differentiated to give the joint velocity equation (A2.2). This is differentiated again to give the joint acceleration equation (A2.3). If there are N+1 samples (including the initial sample 0) during time t_f, the sampling rate will be N/t_f.

Joint displacement:

$$q(t) = a_0 + a_1 t + a_2 t^2 + a_3 t^3 + a_4 t^4 + a_5 t^5 \qquad \text{(A2.1)}$$

Joint velocity:

$$\dot{q}(t) = a_1 + 2a_2 t + 3a_3 t^2 + 4a_4 t^3 + 5a_5 t^4 \qquad \text{(A2.2)}$$

Joint acceleration:

$$\ddot{q}(t) = 2a_2 + 6a_3 t + 12a_4 t^2 + 20a_5 t^3 \qquad \text{(A2.3)}$$

Where: $q = \Theta$ (for revolute joints) or d (for prismatic joints), from the DH notation (Denavit and Hartenburg 1955).

The discrete time at sample n (= {0,1, ..., N}) is: $t = \dfrac{n \times t_f}{N}$ (A2.4)

The polynomial contains six unknowns and they are found from the six boundary conditions:

$q(0)$ = Initial joint displacement,

$\dot{q}(0)$ = Initial joint velocity,

$\ddot{q}(0)$ = Initial joint acceleration,

$q(t_f)$ = Final joint displacement,

$\dot{q}(t_f)$ = Final joint velocity, and

$\ddot{q}(t_f)$ = Final joint acceleration.

The coefficients are therefore:

$$a_0 = q(0) \quad \text{(A2.5)}$$

$$a_1 = \dot{q}(0) \quad \text{(A2.6)}$$

$$a_2 = 2\ddot{q}(0) \quad \text{(A2.7)}$$

$$a_3 = \left(20q(t_f) - 20q(0) - (8\dot{q}(t_f) + 12\dot{q}(0))t_f - (3\ddot{q}(0) - \ddot{q}(t_f))t_f^2\right)/\left(2t_f^3\right) \quad \text{(A2.8)}$$

$$a_4 = \left(30q(t_f) - 30q(0) + (14\dot{q}(t_f) + 16\dot{q}(0))t_f - (3\ddot{q}(0) - 2\ddot{q}(t_f))t_f^2\right)/\left(2t_f^4\right) \quad \text{(A2.9)}$$

$$a_5 = \left(12q(t_f) - 12q(0) - \left(6\dot{q}(t_f) + 12\dot{6}q(0)\right)t_f - (\ddot{q}(0) - \ddot{q}(t_f))t_f^2\right)/\left(2t_f^5\right) \quad \text{(A2.10)}$$

INDEX